EXPERIMENTS IN PRACTICE

History and Philosophy of Technoscience

EXPERIMENTS IN PRACTICE

BY

Astrid Schwarz

Routledge
Taylor & Francis Group

LONDON AND NEW YORK

First published 2014 by Pickering & Chatto (Publishers) Limited

2 Park Square, Milton Park, Abingdon, Oxon OX14 4RN
711 Third Avenue, New York, NY 10017, USA

Routledge is an imprint of the Taylor & Francis Group, an informa business

First issued in paperback 2016

BRITISH LIBRARY CATALOGUING IN PUBLICATION DATA

Schwarz, Astrid E., 1963– author.
Experiments in practice. – (History and philosophy of technoscience)
1. Science – Philosophy. 2. Science – Methodology.
I. Title II. Series
501-dc23

ISBN-13: 978-1-8489-3485-6 (hbk)
ISBN-13: 978-1-138-70637-8 (pbk)

Typeset by Pickering & Chatto (Publishers) Limited

CONTENTS

LIST OF FIGURES

INTRODUCTION:
TOWARDS AN EXPERIMENTAL MODE IN SCIENCE, SOCIETY AND PHILOSOPHY

For a long time experimentation was perceived as a scientific method that is implemented primarily within the walls of a laboratory. This literal as well as conceptual boundary has since been exceeded: nowadays, experiments are happening everywhere and they might be conducted by just about anybody. Virtually any social, political, artistic or economic action can be conceived as an experiment, and citizens can signal their willingness (or refusal) to be part of an experiment in a variety of ways, be it at the level of society or in questions of personal lifestyle. Evidence of the presence of the experimental mode in society is compelling. Artists make use of experimental techniques ranging from genetic engineering to ambient intelligence environments; scientists and engineers propose large-scale experimentation in order to prevent or at least slow down climate change; in political science and public administration, new policies are framed as experiments in governance or collective experimentation; and political movements such as the ones observed most recently in the Middle East are described as experiments in democracy. Finally, biographies and identities are seen as tentative and provisional, a perspective echoed in the new NBC pilot comedy series *My Life as an Experiment*.

Evidently, the experiment has become a widely accepted cognitive strategy in the arts and sciences, in technology and politics, in education and in practices of the self. The idea of experimentation has come to be associated empathetically with the notion of the so-called knowledge societies. Here, a variety of conceptions serve to designate the experimental constellations of learning and action. They include real-life or real-world experiments, real-world simulation, experimental installation and innovation, adaptive or experimental management, social or collective experimentation. Others, such as computational simulation and prototyping, designate experimental design processes. All these designations indicate that experimentation is currently a dominant mode of experiencing, describing and designing cultural as well as natural environments.

This study illuminates the experimental mode from a variety of philosophical perspectives. Epistemological problems such as the interaction between deductive and inductive methods, the balance between 'thoughts and things' and the heuristics of 'trial and error' are one part of the story. Another has to do with questions about the historical genesis of the experiment as a form of experience, about the heuristic role that was (or was not) accorded to the experiment over time, about the contribution of the Baconian sciences and, finally, about the work of purification required to distil *experientia*, *experimentum* and *observatio* into separate phenomena. A third narrative strand is the normative perspective that considers how risk and surprise are perceived and negotiated, what epistemic and non-epistemic values are involved and how economic, media, political or ethical concerns intersect in experiments. Natural philosophy, of course, is another important strand related to the previous three perspectives, as conceptions of nature change the way we act upon nature, and vice versa.[1] All this comes together when we consider how experiments are designed and when they begin and end. The location of an experiment – where and how nature is put on display – is a question of scientific practice and methodology. Experiments in a clean lab are radically different from experiments in the field, while biological field experiments differ again from social experimentation.[2] An analytical framework is offered that spans the gap between a lab ideal and a field ideal, making it possible to elaborate on the diversity of experimental strategies in a more systematic way.

Historically speaking, the overwhelmingly positive reception of experimentation is a fairly recent phenomenon. For the most part the experiment appeared as an ambiguous phenomenon. It was regularly identified with the dark side of modernity: knowledge achieved through experimentation was considered to be ideologically charged. Until well into the twentieth century, such sceptical attitudes were quite common; the experiment stood for a purely scientific procedure that speaks truth to power about nature – including human nature. This was expressed most powerfully by Hannah Arendt when she characterized concentration camps as 'laboratories in the experiment of total domination' of human nature.[3] Meanwhile, the metaphor of nuclear weapons as an experiment that humanity is conducting on its very survival was rejected by Günther Anders for its trivialization of the issue.[4]

Contested Environment

During the 1950s voices disparaging traditional scientific knowledge production as reductionist and destructive were raised more and more forcefully in Western societies. Environmental damage and degradation were now seen as a consequence of technological hubris and an ideology of scientific progress directed against nature. These critical issues, motivated by political, aesthetic and philosophical reasoning, crystallized in the emerging environmental movement, which prepared the ground for a cultural shift in which 'the environment' came to be regarded as a problem.[5] Certain empirical phenomena, such as the unintended effects of certain technologies on humans or the environment, were no longer regarded as acceptable consequences of modernization but were seen instead as part of an overarching crisis: the air, water and the earth were now 'polluted'. However, although the dominant political narratives about 'the environment' expressed suspicion of the knowledge produced by 'science and technology', they turned to a particular scientific field in the expectation that it would provide an 'alternative' way of knowing nature, given that it was considered to be a 'soft science'. This scientific field was ecology, which at that time was an academically diffuse science. Ecology was to promise a different, more harmonious and holistic scientific approach and thus to provide a more benign, less harmful way of knowing nature.

This expectation was foiled for many reasons, one of them being that ecological research programmes, despite tackling complexity, were also constrained by reductionist approaches. Ecology too offers a form of practical knowledge (*Verfügungswissen*) that tends to model nature just as molecular biology or the physical sciences do. Ludwig Trepl, a historian and theorist of ecology, makes this point when he emphasizes that ecology rigorously models nature in accordance with its epistemic values, that is, in such a way that nature conservation pays off.[6] Keeping this ecological rationality in mind, it is perhaps not so surprising that ecology became involved in huge technoscientific programmes such as the US space programme, involving, for example, a sub-discipline called cabin ecology. In this context, ecology contributed as much as any other science towards modernist experimentation and the competitive race to put man on the moon.

And yet the misplaced expectation in ecology as an alternative, benign science initiated a productive dynamic. The political claims vis-à-vis scientific ecology and the disappointment on both sides regarding the limits of communication proved to be indicative of the changing conditions of scientific research in society. A debate between science and its public on the methods and models to choose and on how to act in a time of crisis fostered new modes of legitimizing scientific knowledge production.[7] Particularly from the 1980s onwards, this debate intensified the spread of experimental approaches beyond the walls of the laboratory. These new constellations of science and politics, involving an acknowledgement that environmental problems are at once political, ethical and scientific problems, fostered a trend in which 'experimentation' became an ever more polymorphous concept, culminating in the successful narrative it presents today. Accordingly, some have argued that scientific research has increasingly become an agent of social change and that it therefore calls for new forms of legitimacy, information and participation.[8] Each of these observations serves to characterize the so-called knowledge society.[9]

In the following, two aspects of this all-encompassing analysis of the new configuration of science and society will be emphasized. The first concerns the way the environment has been transformed by changes in the way science looks at it. The second considers how the environment has become a different object as a result of changes in the way it is interpreted by politics and culture. Since neither of these transformative aspects can be described in isolation from the other, the philosophical implications of this shift are considered with a view to epistemological as well as historical questions. In particular, two claims are discussed in this context. If science is increasingly accepted as an agent of societal change, then parts of science are turned into a field experiment. Second, the call for public participation to shape and control knowledge production draws at least some parts of science into a democratic process.[10] In both cases, the suggestion is that the new contract between science and society is being forged in an experimental mode.

Turning Science into a Democratic Endeavour

Given the prominence (until recently) of a sceptical attitude towards the experimental mode, it is interesting if not surprising that the current mode of experimentalism encounters hardly any criticism. Even the highly contentious, historically and morally problematic view of citizens as guinea pigs in collective experiments goes largely unnoticed and unchallenged. This may be due to the fact that the framing of social experimentation has changed fundamentally. Now, the experimentalists who are designing the experiment are themselves the subjects of the experiment. This constellation emerged from attempts to link science-based experimentation with democratic procedures. It came in response to an earlier incarnation of a rather technocratic 'experimental approach to social reform'[11] as promoted by social scientists and, a little later, by US economists in the 1960s as well.[12] The new ideal society was characterized by the growing importance of standardization and quantification, a kind of mechanical objectivity that simultaneously stood for a new style of government that put its trust in numbers rather than in individuals. It is during this era that the implementation of rigorous experimental and quasi-experimental set-ups was advanced in the 'real world', in field settings.[13] However, even the softer term 'quasi' experiment could not hide the technocratic style of these experimental designs and their seeming irrefutability, which appeared to be corroborated by 'scientific objectivity'. This technocracy increasingly became a subject of criticism in the 1980s. Experimentation and its associated scientifically rigorous method of trial and error were challenged first and foremost in relation to environmental problems, since the experimental method was regarded as an outcome of modern industrialization. Additionally, though, it was challenged in relation to the governmental decision-making processes that turned society into a laboratory. The ensuing conflicts between different interest groups and the lack of consensus became especially apparent in the context of politics and policy making.

In response to this situation, Ulrich Beck's *Risk Society* posed the questions 'whose experiments are these?' and 'on whose backs are they being conducted?'[14] In an explicit step towards a more positive interpretation of the experimental mode, 'society as a laboratory' became a cipher for 'the social risks of experimental research', emphasizing openness towards learning processes.[15] The focus was now on technological innovation and its procedural conception of risk as being limited to scientifically measurable health and safety risks[16]. These 'real-life experiments' (later re-coined as real-world experiments or again as 'implicit experiments') included the release of GMO plants, the use of the insecticide DDT and the implementation of nuclear power plants. In all these cases society at large is exposed to the dangers of scientific error and technological failure, usually on a global scale. This led social scientists and political theorists to reconceptualize the experimental mode in a more encompassing and democratic manner. The proponents of this revised approach call for a redistribution of responsibilities in order to ensure that different stakeholders in society are included in deliberations over how innovative experimental processes should take place and, indeed, why emerging technologies should be implemented at all.[17] The concept of 'real-world experimentation' entails consideration of these issues, with particular emphasis on 'ecological intervention'.[18]

The recent shift in the semantics of experimentation is related to different ontological settings and epistemic strategies that are optimistically oriented towards the future and performed in open spaces of possibility. Virtually every topic or area of life is described in the experimental mode, whether it be a citizen's identity (European vs US), emerging technologies (nanotechnology, synthetic biology), consumer products (smartphones, hybrid cars), synthetic materials (eco-friendly plastics, smart textiles), family and work (social networks), environmental concerns (carbon offsets, green responsibility) or nature restoration (re-establishment of wetlands, reintroduction of wild species). The defining commitment in all these activities is a commitment to negotiation and participation. In other words, the liberation of the experiment takes place in the context of a continual and open-ended deliberation of the rules of the game. These rules comprise regulative, normative and cognitive aspects, and, as has been pointed out by numerous figures, the process of deliberation needs to be embedded in democratic institutions and supported by an informed and attentive citizenship.[19]

The resulting matrix of arguments and values in any given 'experimental setting' affords a snapshot of a state of deliberation (no matter whether it is regarded as a real-world experiment, as social experimentation, as an experiment in design or as a living lab). These deliberations involve a methodological understanding that brings descriptive and normative as well as heuristic and justificatory elements to the fore.

Revealing Snapshots

A snapshot of a real-world experimental situation reveals a state of deliberation and therefore a configuration or matrix of ontological commitments and political values. The picture only comes into focus when natural and artificial things, fact and fiction, epistemic and ethical values, risk and benefit or commission and omission persist for a period of time.[20] This type of democratically monitored 'picture' is ubiquitous within the frame of research politics – indeed virtually anywhere where scientific and technoscientific knowledge production is involved. Sometimes the 'matrix picture', or snapshot, is accompanied by specific visualizations, one famous example being the first colour photographs of 'our blue planet', shot during the Apollo 8 mission in 1968. They became – and still are – iconic for the limits to growth debate.

Such a snapshot analysis can be done for the experiment *Atomausstieg* (phasing out nuclear power) or 'green nanotechnology' but also, for example, for experiments with 'European identity' in the context of science policy and its institutions. European research policy is often portrayed as a laboratory for the democratic (re-)invention of 'Europe'. The collective experimentation process of building a single European knowledge society involves a collective experiment with new technologies.[21] How this tension plays out has been described in relation to a particular European institution, namely, the European Environment Agency. It has been shown – unsurprisingly – that this institution reflects different notions of society as well as diverse cultural dimensions of environmental policy.[22] They argue that civil society is called upon to take a more active role in articulating public values and at the same time providing back-up for administrative institutions, acknowledging ignorance and uncertainty, shifting the deliberation of technologies from downstream considerations of calculable and manageable risks to upstream negotiation of societal needs, and displacing expert rulings with integrative precautionary approaches. Evidence abounds not only that European societies perceive and describe themselves as being in a mode of experimentation but also that they consider it desirable to do so. One such piece of evidence is a report commissioned by the European Commission (2007) entitled *Science and Governance: Taking European Knowledge Society Seriously*.

Experiments in Practice: Conquering the Field

Today's comprehensive, democratically oriented and positively valued experimentalism has led us to believe that the real-life (or social) experiment provides an insight into conceptions of nature, identity, ontology and political rationality. Putting the experiment to work thus affords a form of analysis that can be projected back in time, thereby providing a historically and philosophically contextualized view of an experimentalism that has transformed nature and society.

It is thus a rather broad conceptual map of experimental modes – veritable experimental cultures – that will be explored in the following from a philosophical perspective. This study traces the gradual diffusion of the experiment into society, eventually arriving at our own culture of experimentation today. The snapshots can be taken as a series of self-images in which society is conceived as a kind of workshop where people, things and symbols along with their interrelations of power, trust and responsibility are constantly subject to deliberation.

In the following, particular attention is given to those experimental settings in science that have already transcended the isolated and protected space of the laboratory – the laboratory ideal – or are about to do so. These experiments might be performed in toto in the field or they might be located epistemologically at the interface between the real world 'out there' and the artificial world of the lab. In this interstitial experimental mode it is common for objects, instruments and people to commute between different places while the entities involved are stabilized by these very dynamic interactions. Many, if not most, sciences share experimental configurations like this, such that the distinction between 'applied' and 'fundamental/pure' research becomes confusing if not untenable. Accordingly, many interdisciplinary and even disciplinary fields of knowledge exhibit blurred and unruly conditions but are nevertheless successful in the sense of knowledge production, that is, in the development of research programmes and scientific or technoscientific objects. Pertinent examples include geology and geoengineering, sociology and psychology, chemistry, pharmacy and the medical sciences, the bio- and ecosciences and bio- and ecotechnologies.

In this book, the widespread occurrence and use of 'ecology' as well as its influence and imputed virtue – indeed even the unease and complaints articulated around its success – are considered to be a significant sign of the ambiguities and changes associated with the conceptual expansion of the experiment. Thus it follows that 'ecology' is extremely well suited as a complementary term to 'experiment' in the changing matrix of values and commitments that have come to define the relations between science and society since the nineteenth century. Ecology (or 'border biology', as it has been called[23]) has been from the very beginning an experimental, 'dappled' field science that hovers between lab and field methods, observational and experimental practices, between academies, advocacy and mixed institutions.[24] It is a scientific field driven at once by an inner dynamic and by external factors. Ecological knowledge turns out to be framed as scientific ecology or 'ecoscience' on the one hand and as 'ecotechnology' on the other.[25]

In the following, philosophical questions are raised about the epistemic and ontological qualities of the experimental mode in environmental matters, which provides the basis on which to configure and reconfigure laboratory and field, science and society, the artificial and the natural. Here, the cultural phenomenon of 'greenness' will emerge as a significant cipher for today. It serves as a pivot or organizing principle for collective experimentation with novel technologies; examples of this will be discussed mainly for European and North American societies.

1 (DE)LIBERATING THE EXPERIMENT: A DIFFERENT HISTORY OF THE PHILOSOPHY OF SCIENCE

Bacons Organon soll eigentlich ein heuristisches Hebzeug sein. (Bacon's Organon should really be a heuristic hoist.)[1]

Georg Christoph Lichtenberg

In the 1980s, philosopher Ian Hacking advocated a rethinking of the role of hands and eyes in the construction of scientific knowledge. His argument for the importance of experimentation was accompanied by the hope that it might launch a 'back-to-Bacon movement, in which we attend more seriously to experimental science'.[2] Now, some three decades later, one has to acknowledge that Hacking has been tremendously successful, even if he was not the first philosopher of science to point to the importance of the Baconian conception of experience through experimentation, which has been referred to sympathetically and critically by Karl Popper, Thomas S. Kuhn and Paul Feyerabend, for example. However, nowadays Francis Bacon is undoubtedly the historical witness most often called upon when describing recent transformations in the production of knowledge and its epistemological and ontological status. Witness, for instance, the expression 'the current state of human affairs in the wake of our collective Baconian transformation', or the writing of another author who reads Bacon as a precursor to Latour.[3] The renewed interest in Bacon is one of the symptoms of contemporary postmodernism, argues Paul Forman, where ends regain primacy and had been ascribed to the scientific methods 'always and everywhere' prior to the Enlightenment, with its high valuation of science, being 'modernity's prime exemplar of progress through reliance upon a proper means'.[4]

To be sure, Bacon has become a lauded philosophical figure for a range of claims he made concerning the origins of experimental philosophy, the equality of applied and pure sciences, and the entanglement of politics, society and science. Last but not least, Bacon is cast as the founder of the inductive method and sometimes as an advocate of inductive logic. This range of approaches suggests that there are rich and fascinatingly diverse readings of Bacon which also have a historical dimension: viewing Bacon as a representative of an inductive science was much more pronounced in the nineteenth century than it is today. Testimony to this is given especially by William Whewell's widely acclaimed *Novum Organum Renovatum*, in which – as the title itself indicates – he critically investigates Bacon's writing in order to sharpen the inductive method. At the same time he advocates advancing a philosophy of science that suggests a balance between hand and mind, between 'thoughts and things',[5] reflecting and modifying what today is referred to mainly as the Baconian tradition.

Multiple Claims on Bacon

Philosopher Madeline Muntersbjorn identifies four different '-isms' that are indicative of how Bacon's work has been perceived over time: Bacon as a naive empiricist, as a pioneering inductivist, as an innovative experimentalist and, finally, as a prescient constructivist. Her clear-cut conceptual scheme serves a particular purpose. Muntersbjorn suggests that not only is it valuable to critically assess the different philosophical positions that have been developed during the history of Bacon's reception but that this plurality is helpful in enabling us to 'recognize the mutually sustaining and jointly necessary character of multiple methodologies within the scientific enterprise'.[6]

In the following sections this proposition is developed and extended by suggesting that all four labels can contribute to a better understanding of Bacon, if each of them is used as a kind of magnifying glass to identify and query the critical points in each of the others. It is the ambition of this chapter to use this mutual 'perspectivization' to gain a detailed insight into Bacon's ambiguous and seemingly inconsistent philosophy and, in so doing, to highlight the role of Bacon as an innovative experimentalist inside and outside the walls of the laboratory – or rather, outside the alchemist chamber. How do we come to experience at all, and what about the extension of experience through instruments?

In Bacon's philosophy the necessity of learning how to learn in order to know is the key driving force, and he offers various innovative tools to bridge the tension between reliable systematicity and indefinite openness. With this necessary ambiguity in mind the literary forms of aphorism and metaphor will be discussed as well as Bacon's focus on 'written experience' and his development of recording instruments, or tools for inscription.

The following snapshot of the Baconian conception of the recording techniques and rhetoric of scientific experience also constitutes an introduction to a much larger topic and is thus guided by three interests: first, to acknowledge the early conditions in which the sciences emerged, long before the institutional divide into lab and field sciences was established; second, to better understand the role of experiment and induction in the systematization of experience; and third, to see whether the history of the philosophy of science takes on a different shape when it is complemented by these perspectives.

Sir Francis Bacon: A Statesman doing Natural Philosophy

In the history of the philosophy of science two works by Bacon have received special attention: *The New Atlantis* (1627) and what is referred to as the *Novum Organum* (1620). Both of these books are concerned mainly with questions about natural philosophy, about knowledge in what might be called theoretical and applied sciences, and about epistemological, ontological and theological questions. One might perhaps argue that the utopia *New Atlantis* is a literary experiment, with ideas developed methodologically in the *New Organon*. However, it seems important to keep in mind that Sir Francis Bacon, Baron of Verulam, Viscount St Albans, was not primarily a philosopher but was trained as a lawyer, carving out an impressive political career while spending a few years as an advisor to Queen Elizabeth I and then as Lord Chancellor for King James I. The *Novum Organon* was published when he held this position, and so accordingly it is the 'serenissimae Majestati tuae Servus Devinctissimus et Devotissimus, Franciscus Verulam, Cancellarius' who asks the 'wisest and most learned of Kings' to tackle this '*Regeneration* and *Instauration* of the sciences' and to take charge of:

> the collecting and perfecting of a true and rigorous natural and experimental history which ... may be designed for the building up of philosophy ... so that at last ... philosophy and the sciences may be no longer an airy and floating fabric but a solid construction resting on the firm foundations of well weighed experience of every kind.[7]

Bacon's background as a lawyer might have been the reason why he associated a firm foundation for the sciences with the idea of testing evidence in a public venue such as a court. Like the delinquent in court, nature was to be investigated by a methodology that allowed access to and control of truth and knowledge. The information gathered in this manner was to be explained by an elite cadre of interpreters. Despite this tribute to a clearly hierarchical control of knowledge, his method, both at the court and in his scientific work, appeared to be more democratic than earlier scholastic methods.

Bacon's juridical background is often put forward as an explanation not only for his interest in particular methodological and institutional issues but also for the literary form he chose to use in the *Novum Organon*.[8] In his detailed study *Laws of Men and Laws of Nature: A History of Scientific Expert Testimony in England and America*,[9] Tal Golan shows how from the late eighteenth century onwards the link between the laws of nature in natural philosophy and the laws of men in the courtroom remains a close one as regards methods for gathering and testing evidence. Indeed the concept of *fact*, as Barbara J. Shapiro shows, did not originate in natural science but in legal discourse. In her book *A Culture of Fact*,[10] Shapiro traces the fact's changing identity and conceptualization, starting in sixteenth-century England. She identifies the developments in news reporting and travel writing as a crucial step in the transition from the fact as an alleged human action to a proven natural (or human) incident. It was only then that 'the fact' which relies on eyewitnesses and testimony in the court of law came to be absorbed by an experimental philosophy. Thanks to the lawyer and philosopher Francis Bacon, the witness (and therefore the 'witnessed fact') became crucially important in both scientific observation and experimentation. Bacon believed that in the court 'ideas should be controlled by those whose position as custodians of the peace and security of the commonwealth best assured their credibility'.[11] Similarly, the 'Idols', the prejudices and preconceived ideas through which human beings observe the world, also have to be supervised as one investigates small bits of nature in a controlled environment and then requires explanation from skilled interpreters. Accordingly, Bacon viewed treason and heresy as similar to the Idols: all of them had to be unmasked, subordinated and controlled – if necessary also by means of torture, which he considered a reliable means to obtain evidence.[12] This procedure, whereby truth is discerned by a hierarchy of knowledge, and Bacon's intimacy with executive power distinguish him as a 'courtly philosopher'. Bacon's role in reinforcing the king's sovereignty, and therefore the monarchic state, extended also to the mercantile system which, with its method of double-entry bookkeeping, not only offered an inscription device and eventually facilitated the manufacturing of facts but also supported a mode of government, as Mary Poovey suggests in her *History of the Modern Fact*: 'In this embryonic variant of the modern fact, then, we see particulars harnessed to a general claim, whose acceptance required believing that the precision of the formal system signalled virtue itself'.[13]

Historians Andrew Ede and Lesley Cormack regard Bacon as a philosopher who is interested in theoretical models; however, at the same time they point out that Bacon was deeply interested in the practical application of knowledge and also grappled with the rhetoric of utility, just like his contemporary Galilei Galileo.[14] With their ambiguous conclusion, Ede and Cormack confirm what has been argued above and what can be considered the quintessence of Baconian philosophy: its experimental character in form and content, and the systematic incompleteness of many parts of his work dealing with natural philosophy.

Experimenting with Form

It was sociologist Wolfgang Krohn who argued that Bacon must have known that the plan of the *Great Renewal* was unworkable: it seems obvious that it could neither be managed by one person alone nor surveyed in its full extent. Bacon wanted his philosophy to be 'not only a science of words (*Wortwissenschaft*) but also an experimental science'.[15] Thus a *New Organon* was to provide a toolbox for the *Great Renewal* that included not only a collection of as many known, documented facts from literature as possible but also primarily new facts that were to be extracted from nature by experiment and observation. This incompleteness in form and content presented a challenge to realizing the Baconian project of a permanent reconfiguration of experience, things and axioms. The reader is challenged to work out a sound strategy for interpreting the bits of text and the overall scheme of the collection of texts – in a manner analogous to that of the naturalist or scientist who is asked to search for a *middle path* between collecting bits of nature, questioning the Idols, and scrutinizing the axioms. One might say that the experimental approach that tests theories propels the Baconian project itself. However, this middle path forms the backbone of Bacon's entire programme of how research should be undertaken: *interpretatio* and *operatio*, *facere* and *intelligere*, *scientia* and *potentia* are the guidelines of this programme. It is a programme aimed at conducting pure *and* applied research, engaging in practice *and* theory, and achieving knowledge *and* power. In fact, the slogan 'knowledge is power', so often attributed to Bacon, is misleading, at least if knowledge is understood as power over nature. Instead, Bacon emphasizes that both human science and human power refer to each other and in the end come together: 'Human knowledge and power come to the same thing'.[16]

It is symptomatic of Bacon's conceptual framing that he places emphasis on the predominance of practical over intellectual skills: 'Neither bare hand, nor unaided intellect counts for much; for the business is done with instruments and aids, which are no less necessary to the intellect than to the hand'.[17] Moreover, 'what is most useful in operating is most true in knowing'.[18] Because Bacon accords so much importance to the hand, Krohn stresses quite appropriately that in his work the 'traditional tension between perception and concept is replaced by the one between practice and concept'.[19] This tension between practice and concept appears in the inescapable dependency on *finding* and *inventing* things, and it renders Bacon especially interesting for recent philosophical discussions that are focused on conceptions such as 'thing knowledge' or 'technoscientific things'.[20] Bacon's middle path is not only a recurrent concept in his work but also an epistemological method. As such it raises not only questions for philosophy of science but also for natural philosophy; furthermore, the problem of finding and inventing things requires not only epistemological answers but also ontological ones.

The Vision of a Learning Society

The *New Organon* is a philosophical treatise that was to provide a methodological infrastructure for a renewal of philosophy. This treatise contributes to the *Instauratio Magna* and thus instigates the programme of a 'Great Instauration' or 'Great Renewal'. It follows upon an earlier work that was published as early as 1605: *The Two Bookes of Francis Bacon. Of the Proficience and Advancement of Learning, Divine and Humane* traditionally known as *The Advancement of Learning*. This was Bacon's principal philosophical work in English, and it 'announces Bacon's comprehensive project to restore and advance learning'.[21] There is no doubt that Bacon had a strong vision of learning, of what should be learned, of how and where it had to be done and by whom. One might say that the huge ambition behind the *Instauratio Magna* was driven by this passion not only to understand learning but also to implement it as *the* driving force in society. It includes a description of the cognitive process, considers the consequences for a systematic design of knowledge, and gives quite a detailed plan of institutions in order to implement the proper conditions for learning in society: '[M]y hope is, that if my extreme love to learning carrie me too farre, I may obtaine the excuse of affection; for that *It is not granted to man to love, and to bee wise*'.[22]

The complete opus *Instauratio Magna* was to consist of six parts, the *New Organon* being the second and incidentally the only complete one (edited in two books).[23] The fragmentary character of the entire programme as well as the aphoristic form of the *New Organon* itself already mirror in some sense Bacon's belief in the power of the mode of experimentation and therefore also an estimation of the openness of the endeavour of a 'general description of the knowledge *or learning* which the human race at present possesses'.[24] From this it also becomes quite obvious that the impossible idea of a 'general description of science' is to be taken quite literally. This becomes evident when viewing the second along with the proposed third and fourth volumes.

In contrast to a prevalent view, it would be a misconception to consider Bacon as a philosopher who deals primarily with epistemological questions in a *Novum Organum* viewed as a purely theoretical work. Instead, the reader of the *Organon* is confronted with countless examples – in fact about half of the book – that are drawn from the so-called applied sciences. These might be taken as further evidence of his practical perspective on experimentation and observation. The *New Organon* was to be the new *tool* to control and acculturate nature – in whatever context. Thus the fragmentary character of most of the *Great Renewal*, one could say, represents an experimental habit reflecting Bacon's conviction that however one starts to investigate nature and to arrange the incremental pieces of information in a methodologically comprehensible order (for instance in the form of tables), these acts will give rise to explanatory or predictive power and thus to control of nature.

The title and plan of the third, incomplete part of the *Great Renewal* was to have been *Phenomena of the Universe*. It provides evidence for the argument that the incompleteness of the project was not mainly a deficiency or even a flaw but was rather a necessary expression and, in some ways, representative of Bacon's 'method'. This method consisted in investigatory procedures and in procedures for selecting and arranging already existing information, thus, Bacon's theoretical departure was a dialectic one.[25] The compilation of natural-historical data would include not simply different kinds of natural phenomena – 'a history of the bodies of heaven and the sky, of land and sea, of plants and animals' – but also of available data and descriptions in literature, all the knowledge from existing academic disciplines and from existing histories of crafts and trades. In short, the book was to include all knowledge about nature. Moreover, this would necessarily include all future knowledge about nature. Bacon was convinced that intervention in nature and thus doing experiments would reveal the secrets of nature much better through 'the vexations of art than they do in their usual course' – just as in 'affairs of state we see a man's mettle and the secret sense of his soul and affections better when he is under pressure than at other times'.[26] Only some experiments, to be sure, can have been performed at the time of the writing of an account of the *Phenomena of the Universe*.

The fourth and vaguest part of the collection furthers the present argument for the systematic and intended incompleteness of Bacon's great renewal. It is of special interest here because it was to be an implementation of the method developed in the *New Organon*, namely, to provide an 'intellectual machinery' comprising axioms and observations 'adjusted to the inductive method' and thus arranging and performing the material in a way that would ultimately allow 'mankind unlimited power to control the natural world not by coercion but by complete understanding'.[27] Thus Bacon's suggestion of how the material ought to be arranged and how the intellectual machinery should be used is ultimately aimed at overcoming the dreadful oppression of the natural world. Instead, small bits of nature should be gathered into the controlled world of recording and experimentation and thereby transformed into artefacts. This procedure was designed on the basis of a model from the world of insects, namely, Bacon's famous description of a bee 'which collects its material from the flowers of field and garden, but its special gift is to convert and digest it'.[28] Along with the 'intellectual machinery', this model proves extremely powerful in Bacon's writing. It suggests that the 'true job of philosophy' is not much different from experimenting, and that it is built on a 'closer and purer alliance (not so far achieved) of these two faculties (the experimental and the rational)'.[29]

Fragments and Aphorisms: An Active Form of Learning

Besides the conspicuous and perfectly consistent incompleteness of the *Great Renewal* there is a second formal means of openness, the literary form of aphorisms, which calls for an exploratory reading. Bacon announces and in some sense lays out this literary form in the title of the *Novum Organum*, whose full version continues 'or, Aphorisms of the Interpretation of Nature and Kingdome of Man'. Together with the formal system (fragments/incompleteness) and the dialogue, the aphorism can be considered a literary form that also shaped the philosophy of science as a literary and philosophical genre. After Bacon, philosophers and scientists such as Blaise Pascal, William Whewell, Georg Christoph Lichtenberg, Arthur Schopenhauer, Friedrich Nietzsche and Ludwig Wittgenstein all wrote at least some of their philosophical works in an aphoristic mode. It has been pointed out that aphorisms and scientific facts have certain things in common – both can be regarded as tools in an experimental setting: 'The dramatic appearance of a fact on the stage of an experiment is supposed to have the rhetorical force of dispelling the doubt of sceptics. The aphorism works in a similar way in that it presents a thought solely for its power to unsettle belief or to suggest a possibility'.[30] Sharing a rather eruptive appearance on the scene, fact and aphorism nevertheless differ in their gesture. The fact pursues an experimental situation to its conclusion and does so by becoming allied with theory. As William Whewell put it so aptly, 'we do not see *them* (the facts), we see *through them*'.[31] The corroborated fact is the currency of the learning process that drives the experiment, whereas the emerging fact brings this process to a conclusion and settles it. In contrast, aphorisms invoke possibilities; they open up the arena of ideas and imagination. An aphorism might even be disorienting: aphorisms provoke speculative thinking, including thought experiments. The openness of Bacon's aphoristic style might be looked at as an invitation to get involved in the method he advocates and practices.[32] The reader is urged into an active role while trying to make sense of the ambiguous and apparently contradictory aphorisms. One starts to construct hypotheses and to arrange combinations of possible expressions in order to make sense of the bits of text – just as a natural philosopher ought to do with the small bits of nature that are to be combined in a distinct order and thus step by step to reveal the causes that rule nature.[33] Not only does the reader experience intuitively the openness of the method, but their attention is likewise drawn to the process rather than to the end of knowledge production. Bacon himself defines aphorisms along this line, as short and scattered sentences, not linked together by what he called artificial method. He was fully aware of the heuristic character of the aphorism that displays an unfinished and open form, in the sense of heuristics as an *ars inveniendi*, the art of inventing or discovering (*Erfindungs-*, *Findungskunst*). He suggested that as long as knowledge is formed by aphorisms and observations, it is in growth; but when it once is comprehended in exact methods, it may per chance be further polished and illustrated, and accommodated for use and practice, but it increases no more in bulk and substance.[34]

Bacon appreciates the necessary indeterminacy of the aphoristic mode. The aphorism, and with it the experimental subjunctive, becomes a central literary form in his philosophy, and it was finally the decisive step that also *hoists* Bacon's programme out of the Renaissance context and into Enlightenment thinking.[35]

Bacon himself was aware of the ground-breaking character of his programme. He was the first philosopher who described his own philosophy as a *revolution*, suggesting that his thought might be viewed as analogous to a political overthrow. It has been pointed out that in Bacon's conception of a productive and powerful science, a revolutionary ideal becomes visible that is at once philosophical, moral, cultural and scientific. This ideal can be traced back to Virgil's *Georgics of the Mind*, to which Bacon directly referred, for instance, in his *Advancement*.[36] Here, Bacon's notion of *revolution* goes beyond the contemporary meaning of a circular turning around or revolving of incidences. Instead, it involves another idea of change and another conception of history, one that comes closer to the modern idea of a form of scientific progress that overcomes and overthrows previous social and scientific conditions.

The German physicist and philosopher Georg Christoph Lichtenberg explicitly referred to Bacon as offering a renewal of the scientific method with his emphasis on the experimental and hypothetical character of knowledge. Emphasizing the productivity of the aphoristic form, Lichtenberg suggested that 'Bacon's Organon should really be a heuristic tool'.[37] According to Bacon, the condition of learning, especially of scientific learning, is to have an intellectual tool at hand that makes invention possible. The subjunctive is the perfect linguistic form to express this permanent transgression of the limits of contemporary knowledge. No wonder, then, that Lichtenberg finds support in Bacon's programme for his own 'cardinal scanning formula ... To invent a finder for all things'.[38] It is precisely the *finder*, the '*Tubus Heuristicus*' as he puts it, that invokes the 'sceptical, hypothetical, experimental subjunctive'[39] that is the elementary form of any heuristic methodology in the art of finding. Moreover, the proximity of discovering/finding and inventing things is underscored by the ambiguous character of the finder which is a theoretical as well as practical tool.[40] Both can be identified as typical Baconian elements, and will be explored in detail elsewhere.

Significantly, Bacon not only challenges his readers to endure the uncertainty that attends the openness of his programme but also enrols them as active participants in the implementation of the *Great Renewal*, since they are already contributing to the endeavour just by reading and trying to make sense of it. One might be tempted to take a big leap into the twentieth century and to regard

Popper's aggressive 'heroic and romantic' characterization of a scientist as the ideal reader of Bacon. Popper's scientists are, after all, 'people who are humbly devoted to the search for truth, the growth of our knowledge; people, whose life is an adventure of brave ideas'[41] – in short, people who are involved in an open-ended scientific enterprise and driven by a passion for learning and for the growth of knowledge.

But the openness and dynamics in the formation of knowledge is just one side of the Baconian *Organon*. There is also a strong systematic force that shapes the production of knowledge. Historian Lisa Jardine points out that 'the aphorisms of this method of presentation are not witty, anecdotal sayings' such that the progress of discourse results from the 'cumulative effect of the discrete aphorisms'.[42] Instead, the aphoristic form helps 'to spur the reader on to similar enterprises of his own'[43] so that, as noted above, the readers experience the method themselves and, in doing so, participate in the enterprise.[44]

Thus the central literary form in Bacon's philosophy, the aphorism, and with it the experimental subjunctive, stimulates the reader to engage in an experimental mode. The reader is drawn into a simulation of permanently transforming and transgressing the textual records of what is known; he or she is invited to undertake an open-ended (and speculative) rearrangement and rehearsal of the stated descriptions of things and processes. It is this experimental technique of a feigned or fictitious experiment that is commonly described as a thought experiment; as such, it is an established element in scientific reasoning as well as in literary and philosophical thinking.[45] The thought experiment not only blurs the canonical genres of scientific and literary writing but also unsettles the otherwise rhetorically fiercely defended border between fact and fiction. Bacon's commitment to the aphorism benefits precisely from this imaginative widening of intellectual sensibilities. However, this licence to speculate in the experimental mode is always balanced and restrained by the written form – be it the aphorism or the famous inductive tables.

Inscribing Experience: Practices and Tools

Just like the fragmentary character of his writings and the aphoristic mode, the literary form of metaphor (or what Bacon calls 'written experience') is an important heuristic tool in his open-ended systematic work. The latter is probably the most elaborated formal tool in his methodology and is intimately linked to his method of induction.

Written experience requires the development of a practice of taking and representing records – including making notes and tables. Bacon devotes a great deal of attention to the latter in particular. Taking notes and creating tables structures experience in general and becomes a precondition for scientific work, for making *scientific* discoveries, for accomplishing *scientific* experiments and for developing *scientific* knowledge:

> And yet so far mental effort has had a much more important part to play in dis-
> covering than has writing, and indeed experience has yet to be made literate. And
> no discovery should be sanctioned save that it be put in *writing*. Only when that
> becomes standard practice, with experience at last becoming literate, should we hope
> for better things.[46]

A section in Chapter 3 below will be devoted to this impressively reflected practice. First, though, the role of experience and experiment need to be considered
in the context of the method of induction. Nonetheless, the question of representation should always be kept in mind, if only because all method was for
Bacon 'to a greater or lesser extent an artifice for convincing presentation'.[47]

Experientia, Experimentum

The attempt to translate the Latin *experientia* and *experimentum* seems to yield
a minor difference between the two words. One dictionary offers 'assay, sample'
as a translation of 'experientia' and 'sample, (object of an) assay' but also 'proof'
for 'experimentum'.[48] More semantic diversity comes with the verb *experior*, which
embodies a number of meanings from a wider array of semantic fields. To list just
some of the meanings of *experior*: to gain experience by doing trials, to learn; to
essay, to probe; to prove the nature, power, effort, agency of something; to prove
the success of something, to put something to the test; to apply something, to claim
something; to give something a try, to take a chance; to assert a right; to apply
any remedy (med.) and hope for success.[49] The semantic poverty of the nouns is
suggestive of the proximity of experience and experiment in their original use by
scholars until the late Renaissance and until Latin was generally given up as a scientific language. A closer look at the etymology reveals that the common linguistic
root is *periculum*, meaning 'assay/trial, proof', but also 'record, danger', and most
interestingly 'court case, accusal'.[50] All of these meanings are included in the verb
but do not add anything to the differentiation of the nouns. This is also true for the
eighteenth century, when the conceptualization of experience became of tremen-
dous interest in philosophy in general, and particularly in natural philosophy.[51]
Accordingly, in a German-Latin dictionary dating from 1741, the *Teutsch-lateinis-
ches Wörterbuch*, the meanings given for the two nouns differ more explicitly: for
experientia it says 'Gebrauch, Erfahrung, Einsicht' (application, experience, com-
prehension) and 'Versuch, Prob' (assay, sample/test) for 'experimentum'.

In the Middle Ages, *experimentum* was a specialized philosophical and
medical term, going back to Scholastic commentaries on Aristotelian and
Hippocratian texts in the classical period. It exhibited an etymological continuity with the Aristotelian *empeiria* and the Hippocratic *peira*.[52] In medieval

discourse *experientia* denoted experience in general, whereas *experimentum* denoted particular instances that were to be described, albeit in a generalized and impersonal literary form.[53] Bacon also used *experientia* and *experimentum* almost interchangeably 'both for the unforced observation which we might call experience and for the contrived experience which we might call an experiment'.[54] This notion of experiment is clearly different from today's conception even in light of the polysemic character of the word. Though it is true that the beginning of the Modern Age was marked by an important break substantially induced by Francis Bacon, it involves a new understanding primarily of experience and only to a lesser extent of experiment. Accordingly, the most important difference between Bacon's scientific method and that of his predecessors – and also many successors – is this clear commitment to the new role of experience and its formalization. Form is a causal structure for Bacon: 'For the form of any nature is such that if it be in place the given nature invariably follows'.[55] However, our understanding of the experiment is also modified. The experiment not only demonstrates a conclusion suggested by systematic and deductive reasoning but also generates new knowledge.

The Method of Induction: Forming Experience

Bacon expected that the process itself – the collecting, listing and organizing of empirical data – would lead to an entirely new and unprecedented theory. However, he was fully aware of the momentous significance of the difficult passage from data collection to a new way of thinking and of the profound problem of bridging this gap by means of communication: 'Not is it easy to pass on or to explain what I have in mind, for people will still make sense of things new in themselves in terms of things which are old'.[56] Bacon proposes to address this cognitive impasse, and the attendant linguistic incompatibility, by applying the new method itself. The reflexive step is complemented by the new practice of perception and thus performs the change of experience through experiencing.

> In fact, I have only one simple way left for conveying what I have to say; to lead men
> to the particulars themselves and their succession and order, and get them in turn to
> repudiate their notions for a while, and begin to grow used to things itselves.[57]

Again, one is tempted to say that, as 'a genuine midwife of modernity',[58] Bacon facilitates a modern subject, an autonomous and critical mind. This claim has been made particularly by sociologists of science who have represented Bacon as a forerunner of their discipline, suggesting that he was one of the first thinkers who had a cooperative science-based vision of societal welfare, based most notably on an awareness of the historical relativity of truth.[59] In fact, Bacon's fierce critique of what he called the 'Idols' can be seen as a critique of ideology. The method of 'true *Induction*' was to be the appropriate means of 'restraining the *Idols* and driving them off' in order to protect notions and axioms.[60] All of this serves to underline the structural as well as strategic importance of the method of induction in Bacon's work, and on this basis it becomes almost self-evident why he invested so much in the justification and formal development of this method. A fundamental concern of the *New Organon* is to offer a thoughtful conceptualization of the relation between sensual perception and experience, of natural entities and scientific artefacts. This paves the way to 'a true and perfect axiom of knowing' of which the 'precept is this: *that there be discovered another nature which is convertible with the given nature, but which is nevertheless a limitation of one better known to nature like a true genus*.'[61] The process of learning thus implies a link between theoretical knowledge and practical knowledge, but the crucial point here is that true knowledge about nature cannot be acquired other than by intervention. The *ars inveniendi* allows for a discovery of 'another nature' by means of practical activities. Simultaneously, inventive experimentation provides access to 'given nature', which finally brings about the more general postulates (the axioms or laws). The description of dispositions of natural objects and the creation of an artificial effect is seen to be equivalent – we only know insofar as we can do. This is the core of the programme to which Bacon gave a number of names such as 'experimental philosophy', 'scientia operativa', or 'natural and experimental history' – a programme with an emphasis on new material and not on controversies (dialectics) and not restricted to logical operations. Moreover, a different perspective on the ontology of objects is required when technical agency and natural agency are not separated categorically, as is usually done in scientific epistemologies (at least in most post-Cartesian and Kantian analyses). When Bacon says that 'the form of heat ... is the same thing as the law of heat' and that one should 'never withdraw or abstract from the things themselves or the operative part'[62] this recalls discussions regarding technoscientific objects that are viewed as part of an 'apparatus/world complex',[63] generated in phenomeno-technological processes[64] and analysed in terms of novel ontological categories such as that of affordance.[65]

Human Intellect is Like an Uneven Mirror

Bacon was in no way naive about the reliability of sense data and sensation and thus about the biased conditions of experience:

> in fact all perceptions of sense and mind are built to the scale of man and not the universe. And the human intellect is to the rays of things like an uneven mirror which mingles its own nature with the nature of things, and distorts and stains it.[66]

Even more explicitly, he writes: 'the greatest hindrance and distortion of the human intellect stems from the dullness, inadequacy and unreliability of the senses',[67] and '[the senses] either desert or deceive us'.[68] The use of instruments does not solve the problem of the 'uneven mirror'. On the contrary, Bacon points out that 'instruments to amplify and sharpen the senses do not count for much'.[69]

Two elementary techniques help to make the method of induction not only a good tool to get at novel knowledge but the only truly reliable one that enables the 'intellectual machinery' to 'engage in things humble and transitory'[70] and by doing so to provide it with an operative power. The first technique entails transforming mere sensations into text, thereby securing the sense data, making the smallest bits of possible instances available, transforming them into comparable experiences and storing the written experience in the archival memory.[71] In the second book of the *Organon*, Bacon not only suggests a system of tables intended to provide an intelligible structure but also wastes no time demonstrating his recording technique straight away: instances of a property ('nature') – one of his two examples is heat – are to be collected and entered into a first *'Table of Essence and Presence'*; a second *'Table of Divergence or Absence in Proximity'* then gives the uncertain or negative instances; and the third is the *'Table of Degrees or Comparative Table'*. The tables allow for a panoramic view over a rather wild collection of known instances, but also 'phantasms, mere opinions, ill-defined and notional conclusions, and axioms altered daily'.[72] The tables afford a systematic order by rendering visible the similarities, deficiencies and gaps, they enable rejection and exclusion of instances, which is 'the first job of true *Induction*',[73] and they provide a basis for further investigation and decision making. In this sense they 'perform an experiment or Investigate further'.[74]

One might say that the aphorisms and metaphors are the subjunctive forms in Bacon's epistemology, whereas the tables and the logical procedures induced then can be considered as the indicative forms. The inductive method obviously serves a dual purpose: first, the so-called indicative form is aimed at securing knowledge through a comprehensible procedure that controls and guides hypothetical thinking; to put it in Bacon's language, this is the building of anticipations:

> The mind must surely be prepared and shaped in such a way as not to exceed due degrees of certainty, and yet understand (especially at the start) that what it has done very much depends on what it remains to achieve.[75]

Philosopher Mary Hesse has pointed out that Bacon's concept of anticipation can be regarded as a proto-hypothesis in terms of the role of experiments and inductive forms. As was shown above, both of these do not merely serve as confirmations of the already known but are guiding tools towards an advancement of learning. Thus the *'Table of Divergence or Absence in Proximity'* served to collect the negative cases in order to learn from these uncertain or false instances.

One is tempted to read this emphasis on 'the negative' as foreshadowing one of the most prominent issues in twentieth-century philosophy of science, namely, that concerning the identification of scientific practice with a falsificationist approach, which presupposed the impossibility of inductive inferences. However, Bacon's approach should clearly be associated instead with the critics of Popper's model, especially Thomas S. Kuhn. Like Kuhn, Bacon sees conceptual frameworks as guides to epistemic activities, in which the quality of an epistemic chain from instances to lower axioms, and then from lower to higher axioms, is permanently scrutinized.

Nineteenth-century philosopher William Whewell provides further evidence for this association of Bacon and Kuhn because he adds another 'Kuhnian argument' to the prevalence of the conceptual framework – namely, the mental element that is being formed in a process of disciplinary deliberation. This argument concerns the 'disciplinary matrix' (one of the many synonyms for 'paradigm' used by Kuhn), that is, the role of theory building in the scientific community, with its shared beliefs that include values, instruments and techniques, even metaphysics:

> the Inductive truth is never the mere *sum* of the facts. It is made into something more by the introduction of a new mental element; and the mind, in order to be able to supply this element, must have peculiar endowments and discipline.[76]

'Doing Experiments': Giving Form to Experience

If records and tables are the first technique to give the inductive method some operative power, the literary forms 'fragment' and 'aphorism' can be considered the second technique since, as discussed earlier, they embody the subjunctive which invites an intellectual openness and even speculation. The use of metaphor is ultimately an experimental technique for Bacon's inductive *ars inveniendi*. Metaphors are frequently used by Bacon, but they serve especially as a technique when he treats them as models, such as in the above mentioned bee model used to explain his middle path. The bee model appears in an aphorism using a form of language that approximates an analytical style. Bacon explains that the axioms are to be drawn 'from the sense and particulars by climbing steadily and by degrees so that it reaches the ones of highest generality last of all; and this is the true but still untrodden way'. This true but untrodden way is radically opposed to the widespread conventional way that 'rushes up from the senses and particulars to axioms of the highest generality and, from these principles and their indubitable truth, goes on to infer and discover middle axioms'.[77] His steadily ascending step-by-step method – which is conceptualized as a material device (such as a table) – helps maintain the distinction between certainty and uncertainty regarding particulars. It also helps in deciding whether experiments ought to be done to secure the epistemic ascent from the senses to instances to more general postulates. It is this procedure that is described when Bacon claims that we have to 'ascend the proper ladder by successive, uninterrupted or unbroken steps, from particulars to lower axioms, then to middle ones, each higher than the last until eventually we come to the most general'.[78] The quality of knowledge thus amounts to the quality of an epistemic chain that stretches from sense data to concepts or models, and finally to theories and laws or law-like sentences. The connective power of this chain is that it can be traced in both directions. It thereby secures the connection between the higher axioms and instances and rises above mere speculation. This connective power is strengthened especially by further scientific inquiry and experimentation – they reach up, so to speak, to Bacon's intermediate axioms which are 'not far from bare experience' and which Bacon holds to be of such importance that 'men's fortune and affairs depend' on them. As Bacon points out, 'we should not supply the human intellect with wings but rather with leaden wings to curb all jumping and flying up'.[79] Since experiments can *make* experiences, they can ensure the bond between the lower axioms and the most general ones. And so it is experiments that can play a decisive role in Bacon's method of induction. The experiment is the heuristic tool of an active scientist; it is a planned action in which instances of experience are brought forward, transformed into written experience and finally put on record in theories or entities of higher abstraction such as laws – at least, one is tempted to add in the spirit of Bacon, for the time being.

In fact, Bacon distinguishes between different types of experiments, thereby underscoring the importance he attaches to them. He distinguishes between fruit-bearing and light-bearing experiments, the former being conducted merely by practitioners with a commercial aim. Only the latter type of experiment can help to discover causes and axioms 'while they are of no use in themselves'.[80] Pamela Smith argues that this distinction of the two types of experiment was part of the strategy to distance the new type of philosopher from artisans and practitioners while at the same time adopting the values and goals of the arts.[81]

'Doing experiments' signifies the process of giving form to experience and permanently rebuilding and justifying the connection between finding and inventing things. The Baconian inductive method does not offer a plan for correcting or polishing the 'uneven mirror'. In fact, it is very useful to be aware of the scratches and blind spots in it and, above all, to appreciate it for what it is, namely, a call to constantly improve ourselves and our relationship to our environment. This metaphor of the 'uneven mirror', with its emphasis on the interdependence of finding and inventing things and on the active role of perception, might be revitalized from the point of view of a recent philosophical discussion on the conception of affordance. It offers a theory of dispositional properties, where things are looked at as objects with properties that are useful to their beholders and can thereby become objects of design. For instance, water has the dispositional property to freeze at 0 degrees Centigrade and to form a layer of ice. This layer of ice *affords* skating. Interestingly enough, the concept was first developed by psychologist James Gibson, who characterizes the relation between an observer and his/her environment as a permanent experimental interdependence. 'An affordance points both ways, to the environment and to the observer'.[82] 'Perceiving is a psychosomatic act, not of the mind or of the body but of a living observer'.[83]

Bacon does more than provide us with a historical definition of the scientific experiment, and what 'he called "artificial experiments" became the model for "experiment" tout court by ca 1660'.[84] While the reduced notion of experiment as 'artificial experiment' (under laboratory conditions) may have become common in science, it fits neither with Bacon's pervasive experimentalism nor with today's broad adoption of the experimental mode. Moreover, it is also too narrow in respect to the multifaceted and differentiated accounts of experimentation in recent studies in the philosophy of science. The Baconian *New Organon* draws us in to participate in an experiment with experience and therefore challenges us to think about the conditions of possibility of learning in a fundamental and radical sense. It suggests that the selection of form is decisive in the construction of a heuristic tool and calls for an epistemology that is founded on notions of co-action, co-working and co-habitation, rather than categorical dichotomies. Accordingly, epistemic things are afforded, they are produced by scientists even as they arise naturally: 'As for works man can do nothing except bring natural bodies together or put them asunder; nature does the rest from him'.[85]

2 THE PHILOSOPHY OF INDUCTIVE SCIENCES

Having scrutinized Bacon's methodology of experimentation and appreciating it in the context of his *Instauration Magna*, another snapshot from the history of the experimental sciences takes us to William Whewell (1794–1866) and his rereading of 'Baconian' experimentation and experience in the nineteenth century. In 1858 Whewell published his *Organum Novum Renovatum* as the second part of a larger work on the philosophy of the inductive sciences.[1] Whewell was very sensitive to the task of developing transparent and reflexive conceptions of experience, experiment and observation, and was generally concerned to create a powerful language of science. He supported a reformed type of Baconianism, which basically meant a philosophy of science that is embodied in the practice of a science throughout its history. And just like Bacon, he sought a 'middle way' between radical rationalism and extreme empiricism, criticizing the German idealists on the one hand for their focus on the ideal, a strongly subjective basic principle,[2] and Locke and the Sensationalist School on the other hand for their focus on the empirical, objective basic principle. In pursuit of his middle way Whewell insisted on a two-sided constitution of knowledge, thus embracing a subjective as well as an objective dimension. With his *Fundamental Antithesis* he argued that there are mainly two opposite elements of which knowledge is composed, namely, ideas and sensations, the latter being 'the *Objective*' and 'the Ideas the *Subjective*' part of every act of perception or knowledge'.[3] Philosopher Menachem Fisch points out that '[w]ith the formulation of the Antithesis Whewell takes his place with the great epistemologists. It parts early with traditional empiricism, but it also goes an important step beyond Kant',[4] mainly because Whewell does not subscribe to the notion of fixed forms of intuitions and categories. Instead, his stock of *Fundamental Ideas* is essentially open. Whewell's entire conception of the possibilities of knowing is pervaded by an antithetical structure. In addition to the pairing of sense and idea, there are also fact and theory, and form and matter. Following his belief in a middle way, he was convinced that in perceiving external things, ideas and sensations are very closely connected but, for analytical reasons, have to be separated as well as possible. He described this as follows: 'The antithesis of *Sense* and *Ideas* is the foundation of the Philosophy of Science. No knowledge can exist without the union, no philosophy without the separation, of these two elements'.[5]

As in Bacon's philosophy, a central question in Whewell's philosophy of science is: what does *form* mean or imply, either as causal structure, as visual instrument, or as linguistic or arithmetic expression? Whewell argued that the instance of perception consists in connecting sensual impressions according to abstract relations such as space and time, number, or cause. These kinds of connections can also be contemplated independently of those things to which they are applied, that is, they can be conceptualized in isolation from one particular instance of a perceived thing. Accordingly, the action of the mind is not thought to create the object, though 'it may mould, combine, and interpret it. At most, let us say, it half-creates the object'.[6] This apt expression highlights the dynamics and openness of the necessarily two-sided character of the 'form' which involves both an act of perception and a designation deduced from the 'relations' involved – in short, an active recipient (be it a scientist or a philosopher).

Another opposition Whewell was interested in is the one of theories versus facts, which is closely related to the foregoing one of ideas versus perceptions. Facts are 'particular observations'; and since observations are an activity of the mind, Whewell argues that a fact is a 'combination of our Thoughts with Things in so complete agreement that we do not regard them as separate'.[7] Paraphrasing Whewell, theories are built upon facts, and facts involve theories. Accordingly, we may look at theories as being distinct from facts, because they can be considered as belonging exclusively to the class of ideas. Yet we are not able to separate facts from mere sensations, because they always involve an active part of the mind, namely, theories – or indeed also opinions. This is why Whewell's ultimate conclusion regarding facts is that 'we do not see *them,* we see *through* them'.[8]

Obviously, therefore, the process of learning involves both ideal and empirical elements and requires an active recipient to combine the structures in the mind and the sense data perceived. Accordingly, induction is the process of adding a new element to the existing combination of instances by an act of thought by which '(t)he facts, the impressions on the senses' are 'bound together under the form of laws and principles'.[9] It is this interest in the development of an inductive method that forms the recurrent theme in Whewell's very broad spectrum of interests that ranged from geological to economic and from architectural to religious issues. Small wonder, then, that he attracted scholars from various fields. Whewell corresponded with Charles Darwin, Charles Lyell, Edward Forbes and frequently with Michael Faraday; he also shared a lifelong exchange with some of his fellows from Trinity College in Cambridge, especially with the founders of the '*Philosophical Breakfast Club*': besides Whewell, it included mathematician Charles Babbage, economist Richard Jones and astronomer John Herschel.

In 1831 Whewell wrote to Jones: 'I do not believe the principles of induction can be either taught or learned without many examples'.[10] This emphasis on empirical data and particular instances is all the more convincing as Whewell held a chair for mineralogy at the time and was involved in experimental studies and systematic observation. In this period he also made important contributions to tidal research, initiating a large-scale worldwide project of tidal observations to specify and refine already existing laws using the power of computing and graphical tables. This will be discussed in more detail when we turn to the tension between observation and experiment and, with it, to the distinction between 'experience in nature' and 'experience of nature'.

Bifurcation of the Conception of Experience: Experiment and Observation

In the eighteenth century experimental practices became highly diversified; they were also increasingly reflected upon and thus more firmly grounded in methodology. They became an established form of knowledge production, giving rise to a systematic storage of things and the concomitant founding of places of scientific competence such as museums, botanical gardens and laboratories.[11] It is during this time of diversification and differentiation that the distinction between observation and experiment was established conceptually. It is important to keep in mind that both kinds of experience are pursued in a deliberate fashion: observation does not just happen but is the effect brought about by exploring a situation. Even in Bacon's writing *observationes* was an unforced experience, which reflects the common use of the word at the time, whereas *observatio* was used in a prescriptive sense to denote 'an act guided by a rule, protocol, or code of behavior, and therefore different from *experientia*, which can be of *anybody*'.[12] At about the same time *observatio* emerged as a *philosophical term*, typically in the plural and as part of the expression *experimenta et observationes*. In her careful study of observation as an epistemic genre, Gianna Pomata stresses that the assimilation of the term into philosophical language was a 'very slow business, fully completed only in the eighteenth century'.[13] The hesitant adoption of *observation* into the standard philosophical vocabulary still affects Whewell's writing where, for long stretches, he couples observation with experiment – significantly, however, in the reverse order, 'observation and experiment'. This is a first hint, as will be shown in greater detail below, that Whewell wanted to develop a philosophical language in which scientific observation is of crucial importance. Initial evidence for this is the fact that in his most prominent contribution to the philosophy of science, the *Novum Organon Renovatum*, a significant part of the book deals with 'Methods of Observation'. Together with aphorism VII in chapter IV '*Of the Colligation of Facts*', this provides a first impression of the crucial importance of observation in Whewell's philosophy: 'Science begins with *common* observation of facts; but even at this stage, requires that the observation be precise. Hence the sciences which depend upon space and number were the earliest formed. After common observation, come Scientific *Observation* and *Experiment*'.[14]

But let us step back again into the eighteenth century when the philosophical bifurcation of experiment and observation became established, though most often with an explicit reference to one or the other. In her study *On the Empire of Observation 1600–1800*, Lorraine Daston notes that the two forms of inquiry were regarded by almost all natural philosophers 'as working in tandem'.[15] Encyclopaedist Jean le Rond d'Alembert characterized the interaction between observation and experiment as a never-ending loop, while Joseph Priestley emphasized how experiments derive observations, which in turn inspire new experiments, providing yet more observations – thus both describe a permanent movement back and forth between the two modes of scientific experience driven by curiosity.

Another important source of information about the perception and use of observation and experiment are the eighteenth-century encyclopaedia projects such as Zedler's *Grosses vollständiges Universallexicon aller Wissenschafften und Künste* (*Grand and Complete Universal Encyclopaedia of all Sciences and Arts*). The *Universallexicon* was published in sixty-three volumes between 1732 and 1754, and it presents a strongly inter-relational representation of both concepts – experiment and observation. *Experimentum* is introduced as follows:

> An assay refers to the experience one gains of a thing when it is brought forth by our industry. It stands in contrast to the observation, which is an experience freely given to us by nature ... only observations of this kind are not always sufficient to discover the true nature of a thing ... One must therefore make an effort to see whether one cannot render nature capable of allowing us to see that which serves our learning. This one achieves through the experiment[16]

The difference between the two concepts is drawn mainly along the line of intervening in nature: as soon as we invest in an experiment we make an effort and expend energy on making things visible by using machines or instruments; in doing so, things are isolated and reduced to what might then be identified as a 'true cause'. This is strongly reminiscent of Bacon's concept of the experiment as a planned action in which instances of experience are brought forward and, in several steps, are finally transformed into entities of higher abstraction such as laws.

Conversely, the *Universallexicon*'s entry 'Observation, *Observatio*' refers to the experiment.[17] 'Observation' is learned experience, '*gelehrte(n) Erfahrung*',

> which differs from experiments or assays and is given to stand beside [i.e. support, Tr.] these ... [and is also called] common experience, is nothing other than a truth, recognized by means of the senses, of something which resides in and occurs in par-

ticular things, without this having been brought to light or to realization by applied industry. Or, in short: An *Observation* is a property or circumstance in a particular thing that is noticed solely by means of the senses.[18]

We observe if we do not invest any effort but rather encounter nature in a contemplative mode. This holds true even if a learned experience is guided by certain ideas, possibly even if rules are adopted of what the things might reveal without being forced. Interestingly enough, neither Francis Bacon nor *Baconianism* is mentioned in the sixty-three volumes of the Zedler encyclopaedia – whereas other contemporaries of Bacon are. Consequently, it bears asking whether Bacon was at all important to philosophers and scientists in the seventeenth and eighteenth centuries. This is not the place to investigate this question in detail; however, it is likely that the work of Bacon was met with much more interest by empiricists than by practitioners of the rationalist or idealist philosophy elaborated mainly in the German-speaking world.[19]

Before attending more closely to Whewell's reformation of the Baconian project and the distinction between observation and experiment in his philosophy of the inductive sciences, it is useful to take a closer look at the legacy left by Bacon through the centuries, which was to frame Whewell's intellectual enterprise.

The 'Popularized Bacon' of Nineteenth-Century Britain

In the first half of nineteenth-century Britain, Bacon was considered to be appropriate for inclusion in *The Library of Useful Knowledge*, one of the most important works to popularize scientific knowledge and thereby also an instrument of general education. The *Library* was published under the auspices of the Society for the Diffusion of Knowledge, founded in 1826, whose intention was 'to promote cheap but well-produced publications for the literate working class'.[20] Books published in *The Library of Useful Knowledge* included *Useful and Ornamental Planting* (1832) and *Vegetable Substances: Materials of Manufacturers*, as well as a series on *Natural Philosophy* (from 1829 onwards), starting with an introductory work on *Objects, Advantages, and Pleasures of Science*. Other types of publication were the *Penny Cyclopaedia* (from 1833 onwards) and the *Penny Magazine* (from 1832 onwards), each edition presenting a mixture of themes ranging from natural history to industrial reportage and poetry. These publications were explicitly addressed to 'workingmen', as the editor of this series called its potential readership, and characteristic of it was a 'moral tone and the emphasis on knowledge and social uplift'.[21]

The *Account of Lord Bacon's Novum Organum Scientarium* appeared in 1828 and accordingly was one of the first books published in the *Library*. In its preface entitled 'To the Reader', the editorial committee announces that this series serves to 'bring before the reader the most amusing facts relating to manners and customs, which the more confined limits of histories can hardly admit of'.[22] The following pages contain two biographies, one of Thomas Wolsey, a brilliant priest and influential politician in the Tudor period around 1600, and one entitled the 'Account of the Novum Organon'. The latter one begins enthusiastically:

> Lord Bacon was the first who taught the proper method of studying the sciences: that is he pointed out the way in which we should begin and carry on our pursuit of knowledge, in order to arrive at *truth*[23] ...
>
> The grand principle which characterizes this great work, and by the proper use of which its author proposes the advancement of all kinds of knowledge, is the principle of *Induction*, which means, literally, *a bringing in*; for the plan it unfolds is that of investigating nature, and inquiring after truth, not by reasoning upon mere conjectures about nature's laws and properties, but by *bringing together*, carefully, and patiently, a variety of particular facts and instances; viewing these in all possible lights; and drawing, from a comparison of the whole, some general principle or truth that applies to all.[24]

One of the first things to note here is that Bacon is introduced as a hero of science and, in particular, as a pioneer of the inductive method. This method is taken to be *the* perfect method. On the other hand, he is also criticized and accused of a 'lofty and poetic diction', that is, he is judged to be not really educational at all, although his rhetoric is acknowledged, as offering powerful and also beautiful conceptions. Indeed, in the report published in *The Library of Useful Knowledge*, Bacon is not identified with the role of a rather naive empiricist. However, what is offered to the reader willing to be educated is a reading of Bacon which comes close to the role of the philosopher of the manufacturing and applied sciences:

> While some things, as gunpowder, silk, the compass, sugar, paper, may seem to depend on certain properties to be developed by Nature herself, yet other things, the art of printing, for instance, contains nothing that is not obvious and completely within human power ... Hence a ground of hope that science might be improved was to be drawn, not merely from the consideration of the unknown operations of nature hereafter to be discovered, but from the probable result of transferring, compounding, and variously applying those laws and operations which were already known.[25]

Baconianism in the Nineteenth Century: 'Bankruptcy of Eighteenth-Century Reasonableness'?

In the academic world, Baconian thinking (to the extent that it was identified with the method of induction) was just recovering from the blow that David Hume had inflicted upon it when he questioned it as a valid form of inference. Hume's philosophy 'represents, in a certain sense, a dead end', a 'bankruptcy of eighteenth-century reasonableness', as Bertrand Russell commented pointedly – going on to express his hope that 'something less sceptical than Hume's system may be discoverable'.[26] Before its downfall in the eighteenth century, 'Baconianism' had undergone a transformation that might be described as a purification towards empiricism, its most famous representatives being John Locke and George Berkeley. In a sense their philosophies formed the British answer to the Continental rationalism represented, for example, by Descartes and Spinoza.

While British empiricists struggled to salvage Bacon's inductivism, the French encyclopaedists Denis Diderot and Jean-Baptiste le Rond d'Alembert championed Bacon enthusiastically and dedicated an entry to him entitled 'Baconisme ou Philosophie de Bacon'. They saw the *Advancement of Learning* as a model of their own endeavour to advance knowledge in society by compiling an organon of all kinds of knowledge, including 'written experience'. The whole article is written in an admiring tone and exposes Bacon as the 'précurseur de la Philosophie' marking an epochal break: 'personne, avant le chancelier Bacon, n'avoit connu la Philosophie expérimentale' ('before the chanceller Bacon, nobody was acquainted with experimental philosophy').[27] The *Novum Organum* is highlighted in particular, because Bacon 'y enseigne une Logique nouvelle, dont le principal but est de montrer la manière de faire une bonne induction' ('teaches a new logic, of which the first objective is to show how to do a good induction').[28]

These rather cursory references serve nonetheless to highlight the fact that Francis Bacon was claimed by different – even contradictory – philosophical positions. Hume was no exception to this. He claimed to be an heir of the Baconian tradition, which he juxtaposed to continental philosophy, referring to:

> Lord BACON* and some later philosophers in *England*, who have begun to put the science of man on a new footing, and have engaged the attention, and excited the curiosity of the public. So true it is, that however other nations may rival us in poetry, and excel us in some other agreeable arts, the improvements in reason and philosophy can only be owing to a land of toleration and of liberty.[29]

And yet, despite Hume's praise for Bacon, Whewell's reformation of the Baconian project had to respond to Hume's scepticism regarding induction.

Humean Scepticism: We Cannot Expect any Certain Knowledge from Experiment and Observation

> A scientist and his wife are out for a drive in the country. The wife says, 'Oh, look! Those sheep have been shorn.' 'Yes,' says the scientist. 'On this side.'[30]

It is once again Bertrand Russell who is said to provide a telling and much-cited example of the famous 'inductivist turkey' that was actually a chicken.[31] It illustrates what is known as the Humean *problem of induction*:[32]

Domestic animals expect food when they see the person who usually feeds them. We know that all these rather crude expectations of uniformity are liable to be misleading. The man who has fed the chicken every day throughout its life at last wrings its neck instead, showing that more refined views as to the uniformity of nature would have been useful to the chicken ... The mere fact that something has happened a certain number of times causes animals and men to expect that it will happen again. Thus our instincts certainly cause us to believe that the sun will rise tomorrow, but we may be in no better a position than the chicken which unexpectedly has its neck wrung.[33]

The problem of induction consists in the validity (or otherwise) of the uniformity principle, which is essentially the claim that the future resembles the past. From past experience the inductive chicken develops the expectation that it will always be fed in the morning, obviously a perilous expectation. A sceptical chicken would have known: 'the supposition, *that the future resembles the past*, is not founded on argument of any kind, but is entirely derived from habit'.[34] According to Hume, all inductive reasoning depends on this uniformity principle, which is presupposed and therefore not derived from reason or any process of understanding but corroborated by experience alone. Russell commented on this quite tellingly:

We have therefore to distinguish the fact that past uniformities *cause* expectations as to the future, from the question whether there is any reasonable ground for giving weight to such expectations after the question of their validity has been raised.[35]

In light of this, the results of an inductive argument can only be probable but never certain (as deductive inferences are): every swan I have ever seen has been white; (in all likelihood) the next swan I see will be white. This is obviously a false inference, because black swans do exist and thus might cross our path, and we can certainly know of them because they have been part of a systematic zoological system since the seventeenth century. The proposition 'all swans are white' is derived from the uniformity principle but turns out to be false, although the primary premise that 'every swan I have ever seen has been white' is correct while the secondary premise 'from past observations I can draw conclusions about future ones' is a necessary supposition.

This led Hume to conclude that inductive inference is not rational since, by definition, rational reasoning is (deductively) logical and thus error-free – and inductive inferences can be false. Even more seriously, the uniformity principle as a justification for induction relies on induction. Since no contradiction is implied by nature developing differently than it did in the past, and since the uniformity principle is therefore not a conclusion based on logical or mathematical reasoning, it is from experience alone that we derive the conclusion that the future resembles the past. Accordingly, the uniformity principle as a justification of induction is itself based on induction, which is clearly a case of circularity.[36]

The dilemma seems to be clear: the inductive principle can be neither disproved nor can it be proved by the appeal to experience:

> we must either accept the inductive principle on the ground of its intrinsic evidence, or forgo all justification of our expectations about the future ... all knowledge which, on a basis of experience tells us something about what is not experienced, is based upon belief which experience can neither confirm nor confute, yet which ... appears to be as firmly rooted in us as many of the facts of experience.[37]

These arguments became and still are an uncomfortable problem for the logical foundation of the empirical sciences.

Taming the Chimera by Way of doing 'Careful and Exact Experiment'

This is not the place to elaborate in detail the complex logical discussions generated by the problem of induction, nor to pretend to be offering an answer to a challenge that, according to Russell, 'has still not been adequately met'[38] – a diagnosis that is still shared by most philosophers of science. More important for present purposes is, first, the historical question of how the problem could arise at all and, second, the question of what, according to Hume, experiment and observation might contribute to the quest for certain knowledge. The fact that 'the experiment' was of fundamental interest to him becomes apparent from the title of his main work *Treatise of Human Nature. Being an Attempt to Introduce the Experimental Method of Reasoning into Moral Subjects* (1739).

Like his predecessors, Hume was concerned to identify certain knowledge and to distinguish it as clearly as possible from uncertain knowledge. This is developed principally in the section 'Of Knowledge and Probability' of his *Treatise of Human Nature*, where he claims that any knowledge is uncertain that is obtained from empirical data by inferences that are 'not demonstrative'. An inference is demonstrative (and only then adds to certain knowledge) if it is imposed by direct observation or if it is based on logic and mathematics.[39] Everything else, that is, knowledge of the future, of unobserved instances or isolated fragments of the present and past, is uncertain knowledge. This is what Hume calls probable knowledge, though this entails no reference (as it customarily would nowadays) to the mathematical theory of probability. It is this distinction between demonstrative certainty and probability that is central to Hume's juxtapositions of observation/instance versus idea and of knowledge versus opinion.

> For to me it seems evident, that the essence of the mind being equally unknown to us with that of external bodies, it must be equally impossible to form any notion of its powers and qualities otherwise than from careful and exact experiments, and the observation of those particular effects, which result from its different circumstances

and situations. And tho' we must endeavour to render all our principles as universal as possible, by tracing up our experiments to the utmost, and explaining all effects from the simplest and fewest causes, it is still certain we cannot go beyond experience; and any hypothesis, that pretends to discover the ultimate original qualities of human nature, ought at first to be rejected as presumptuous and chimerical.[40]

Therefore, to build on experience is certainly the only way to achieve knowledge about the external world as well as about the internal word of our own mind and its capacities, even if it is perfectly clear that the universal principles (such as the uniformity principle) can only be achieved by way of approximation.

> If reason determin'd us, it wou'd proceed upon that principle, *that instances, of which we have had no experience, must resemble those, of which we have had experience, and that the course of nature continues always uniformly the same.*[41]

Hume repeatedly contends that the connection between objects is merely a connection between the ideas of those objects. Probable or uncertain knowledge is necessarily a mixture of external data (sense impressions) and ideas, and 'were there no mixture of any impression in our probable reasonings, the conclusion would be entirely chimerical'.[42] The 'careful and exact experiment' along with observation allows the chimera to be tamed. The experiment depends on forming hypotheses and includes other activities of the reasoning mind; at the same time it provides us with sense impressions (and data) about external objects.

Solving the Dilemma: Certain and Probable Knowledge

To solve his dilemma, Hume differentiates between certain and only probable knowledge and, correspondingly, between philosophical relations that are purely ideal and those that are affected by experience of external objects. Accordingly, Hume's seven philosophical relations of resemblance – proportion in quantity or number, degrees in any quality, contrariety, identity, relations of time and place, and causation – are divided into two kinds, those relations that depend only on ideas and those that depend on objects. These latter include the spatio-temporal relations as well as the causal relations, and both of them give only probable knowledge. In fact, even the relations of the first kind are only certain when they are formalized in the languages of algebra or arithmetic which alone can ensure that 'a long chain of reasoning [can carry on] without losing certainty', as Bertrand Russell points out in the chapter of his *History of Western Philosophy* on Hume.[43]

The crucial point for the problem of induction, and for experiment and observation, is that the concept of cause comes down on the side of uncertain or probable knowledge and thus falls into the category of opinion, or probability – and therefore does not constitute actual knowledge at all. Our causal reasoning, Hume maintains, is determined by custom, and it is 'this impression, then, or determination, which *affords* me the idea of necessity'.[44] It is only the repetition of instances which leads us to believe that A causes B. There is no necessary connection between cause and effect as soon as 'causation is stolen from knowledge'.[45] Consequently, there is no need to ensure a necessary connection between a present event and a future one – or equally, a past event and a present one.

If one wants to understand how Hume's problem of induction became possible in the first place, one needs to see that the break with Bacon's epistemology and the 'theft' of reasoning in relation to cause and effect from the old shelter of knowledge had been prepared before the time of Hume, as Michel Foucault and also Ian Hacking, referring to the former, have shown. The following is a brief account of how their story goes. In the first half of the seventeenth century the domain of empirical knowledge came to be reorganized as the logic of representation and the order of signs, along with the order of things, changed dramatically: 'Ce n'était pas la connaissance, mais le language même des choses qui les instaurait dans leur fonction signifiante' ('It was not knowledge, but the language of things itself that justified them in their significant function').[46] It was George Berkeley who first claimed that the

> connection of ideas does not imply the *relation* of cause and effect, but only of *mark* or *sign* with the *thing signified*. The fire which I see is not the cause of the pain I suffer upon my approaching it, but the mark that forewarns me of it.[47]

This means that the things we commonly take for causes, such as the fire, are not really causes: 'The fire is, after all, the efficient cause, but like all efficient causes it is only a sign'.[48] It is this collapse of distance between cause-and-effect and the sign that prepared for the possibility of experience no longer being based on likeness or identity of nature but only on a habitual connection. At this moment the shift from knowledge to opinion and probability is fulfilled: 'A la connaissance qui devinait, *au hasard*, des signes absolus et plus ancien qu'elle, s'est substitué un réseau de signes bâti pas à pas par la connaissance du probable. Hume est devenu possible' ('The knowledge that divined, *at random*, signs that were absolute and older than itself has been replaced by a network of signs built up step by step in accordance with a knowledge of what is probable. Hume has become possible').[49] What is experienced is a sign; this sign results from the cognitive act of knowing itself, and it is the iteration of this act that – as Hume holds – affords the idea of necessity and consequently establishes a cause. It is only with this shift from causes as the subject of demonstration (thus knowledge) to causes as the subject of a 'network of signs' (within probable reasoning) that Hume's problem of induction has become possible. This shift completes the 'historical transformation by which the signs of the low sciences [such as alchemy] became identical with the causes of the high ... Causes are signs, but the signs suggest the things signified "only by an habitual connection"'.[50] This is what renders Hume's coup against inductive logic so striking: it allows him to make the claim that the connection between cause and effect is no more than habit and custom – or, to invoke Hume's wording again: our reasoning is determined by custom that 'affords me the idea of necessity'. It is only then that induction is converted into the analytical problem we are familiar with today, the one formulated thus: if induction cannot be justified by either matters of fact or relations of ideas, then induction cannot be justified at all.

What follows from this for the role of experiment in empirical research? Is the close link between sign and thing also a step towards a more explicitly experimental philosophy? If it is the case that knowledge is no longer identified with 'absolute' signs given by a divine power, but is released from this kind of certainty into a probing with probable knowledge – does this imply that the experimental mode is empowered because it gives form to knowledge that arises from a world of signs?

3 WHEWELL'S INNOVATION THROUGH RENEWAL

Where it addresses, where it invokes 'truth', thought relativizes this criterion in the moment in which it adverts to it. There is no escape from this dialectical circularity.[1]

George Steiner

In light of the 'bankruptcy' of the logic of induction and, more generally, the contentious character of the Baconian programme in the nineteenth century, it is all the more remarkable that William Whewell and his contemporaries John Herschel and Richard Jones of the 'Philosophical Breakfast Club' of Trinity College, Cambridge, were convinced that a renewed kind of Baconianism would prove to be an adequate method to describe scientific knowledge. The *Novum Organon* was one of their favourite readings even during their undergraduate days.[2] An inductive method, they thought, would be a powerful tool not only to describe and assess the logic of already existing and secured knowledge – that is, to justify the operations, relations and conclusions of received and accepted science – but above all to help foster discovery and generate new knowledge. One of the most important Baconian precepts upon which Whewell drew was the idea that any valid philosophy of science must be developed from actual scientific practice.[3] While Bacon himself could not yet draw on much of a history of modern science, this became an important constraint for Whewell. Although it remains debatable, of course, how consistently and successfully Whewell applied this precept, nonetheless 'he shows us in his works – through numerous apt examples – that his philosophy has been embodied in the practice of science throughout its history'.[4]

Whewell's epistemology embraced inductive as well as deductive elements and he deliberately constructed it *antithetically*, implying that theory and fact, thoughts and things, necessary and experiential truths, language and marks are linked in a necessary relation. Indeed, to a certain extent they are inseparably entangled such that knowledge is of a dual nature. In the end, his theory of the antithetical character of scientific knowledge became transformed into a 'theory of the *sources*' of knowledge, as philosopher Menachem Fisch[5] has pointed out, and therefore went beyond purely analytical claims towards a more comprehensive theory that does not shy away from natural philosophy.

> In every act of my knowledge, there must be concerned the things whereof I know, and thoughts of me who knows ... If we *think* of any *thing*, we must recognize the existence both of thoughts and of things. *The fundamental antithesis of philosophy is an antithesis of inseparable elements.*[6]

This epistemology evident in a *Fundamental Antithesis of Philosophy* and the central role of conceptual schemes in scientific discovery will be explored in the following, with a special attention to Whewell's conception of facts, of scientific objects, and of works in the applied sciences and arts. This will be followed by an example from his own scientific activities, which draws our attention to Whewell as a field scientist and to his concern to determine what constitutes *observation* and *experiment* in the empirical sciences and to distinguish them from one another. It turns out that Whewell's philosophy is a rich source for a strand of thinking that has recently assumed some prominence in the philosophy of science, namely, that which addresses the interaction between hand and mind and, more generally, the importance of manipulating the objects of inquiry through instrumental and experimental intervention.

Stringing Together a Handful of Pearls

As far back as the early 1830s Whewell described induction in his notebooks as a synthesis of the ideal and the empirical which results from both observation and reason. 'Conceptions of the mind' are the intellectual tool that structure perception and therefore are an *a priori* condition for observation:

> Induction agrees with mere Observation in accumulating facts, and with Pure Reason in stating general propositions; but she does more than Observation, inasmuch as she not only collects facts, but catches some connexion or relation among them; and less than Pure Reason ... because she only declares that there are connecting properties, without asserting that they must exist of necessity and in all cases. If we consider the facts of external nature to lie before us like a heap of pearls of various forms and sizes, mere Observation takes up an indiscriminate handful of them; Induction seizes some thread on which a portion of the heap are strung, and binds such threads together.[7]

This metaphor recurs in the form of two key themes in Whewell's method of induction, namely 'colligation' and his general notions or 'conceptions'. Indeed, colligation is the binding together of isolated facts by a general notion or conception, where the latter serves mainly to construe the link between what he calls the *Fundamental Ideas*, *Scientific Ideas* or *Axioms* (such as force) and the phenomena (in this case, purely mechanical phenomena). These fundamental or scientific ideas are not the objects but the laws of thought; they are 'precise and stable' and 'are possessed with clear insight, and employed in a sense rigorously limited, and always identically the same'.[8] The colligation of facts supports an economy of thinking that ideally uses as few conceptions as possible. Every conception represents a connection established 'by an act of the intellect' among every case and 'the phenomena which are presented to our senses. The knowledge of such connexions, accumulated and systematized, is Science'.[9]

Whewell now characterizes the method of induction as a process that is divided into three steps: 'the *Selection of the Idea*, the *Construction of the Conceptions*, and the *Determination of the Magnitudes*', that is, the translation of the conception into mathematical terms.[10] It has been noted that Whewell is not entirely clear about his use of the central terms 'conception', 'induction' and 'colligation', since he either defines one by the other or conflates them in his account. Further, it does not seem evident that much is gained by the term 'conception' above and beyond notions such as concept, rule, method, theory or hypothesis. For example, he holds that conceptions are different from hypotheses, yet at the same time conceptions 'which a true theory requires are very often clothed in a *Hypothesis*'.[11]

Constructing Conceptions and Facts: 'Moving Up and Down a Ladder'

Conceptions rely on either quantitative or qualitative descriptions of phenomena, in other words they differ in regard to their written form and in the kind of symbol selected (name, number, icon). The written form can consist in different inscription systems that might be arithmetic, such as tables with numbers, or diagrammatic, such as schemes, graphs or maps. Whewell's conceptions are thus a central part of the inductive process which 'involves a deliberate imposition upon the data'[12] by connecting and unifying phenomena through the introduction of novel concepts or propositions that finally allow for a selective aggregation of facts – the process Whewell calls colligation (we will come back to this later). This illustrates once again that Whewell was not interested in viewing induction merely as a logical operation or mode of proof or argument but rather as a process of discovery and thus as a heuristic tool for research: 'the inductive step consists in the *suggestion* of a conception not before apparent'.[13] From Whewell's perspective, talking about the logic of induction is synonymous to developing a scheme of the introduced concept as well as the working process to test the concept against the instances. Whewell describes this procedure in terms of moving constantly up and down a ladder. He bolsters his argument by providing a number of rules for the construction of conceptions and yet repeatedly points out that these rules 'must be tested by the facts'.[14] With the ladder metaphor, Whewell brings to mind Bacon's ladder, which was used to illustrate his step-by-step method of proceeding from particulars to lower to middle axioms and, finally, to the most general ones.

Three main questions will be discussed in the following: What exactly are the constructive rules that give rise to a conception? How important are experiments in the process of testing a hypothesis? And how can it be, in Whewell's philosophy of science, that rules are tested by facts even as these facts are so heavily influenced by hypotheses and hypotheses are akin to conceptions?

Finding Conceptions by Means of Constructive Rules

Whewell differentiates between purely quantitative descriptions and those in which qualitative properties play a part. For the latter he formulates two methods or rules, namely, the *method of gradation* and the *method of natural classification*, both of which rely on principles of resemblance (or likeness) that are especially prominent in natural history. It is in natural history, after all, that one encounters most distinctly 'the Ideas of Likeness, of Kind, of subordination of Classes'.[15] Here, Whewell fully appreciates the importance of classificatory systems that rely on names and draw on iconic relations: for the 'division of things into Kinds, and the attribution and use of Names, are processes susceptible of great precision' – as susceptible to precision, indeed, as mathematical operations.[16]

The classificatory transformation from a thing to a kind happens – in botany for instance – by giving each entity (here: organs) a distinct name (an umbel, a catkin, a leaflet) and then assigning each name to a kind, as in the binary system proposed by Linné:

> Linnæus embodied and followed out the convictions which had gradually been accumulating in the breasts of botanists; and by remodelling throughout both the terminology and the nomenclature of botany, produced one of the greatest reforms which ever took place in any science.[17]

Whewell is equally convinced, however, that the mapping of plants and animals should best be done on the basis of numbers, though not necessarily as an arithmetic representation.

> Number supplies the means of measuring other quantities, by the assumption of a *unit* of measure of the appropriate kind: but where nature supplies the unit, number is applicable directly and immediately. Number is an important element in the Classificatory as well as in the Mathematical Sciences.[18]

The number of petals, pistil or stamen can be given to characterize a plant family, say *Rosaceae*, and thus to distinguish different kinds of plants on the basis of these numbers:

the Reform of Linnaeus in classification depended in a great degree on his finding, in the pistils and stamens, a better numerical basis than those before employed ... There are innumerable instances, in all parts of Natural History, of the importance of the observation of number. And in this observation, no instrument, scale or standard is needed, or can be applied; except the scale of natural numbers, expressed either in words or in figures, can be considered as an instrument.[19]

This consideration of natural history is important for our discussion of observation and experiment, since in the context of Whewell's philosophy of induction the giving of a systematic name and a number to a thing that was identified as a plant amounts to using an instrument in the domain of natural history. This applies also to the *Animal Kingdom*, for which Cuvier 'explains, in a very striking manner, how the attempt to connect zoology with anatomy led him, at the same time, to reform the classifications, and to correct the nomenclature of preceding zoologists'.[20] Accordingly, Whewell insists that natural history 'ought to form a part of intellectual education, in order to correct certain prejudices which arise from cultivating the intellect by means of mathematics alone'.[21]

This prestige accorded to natural history and its treatment on a par with the mathematical sciences stems from Whewell's understanding that the inscription of an observation of biological things serves as an instrument provided that it is conducted as an observation of (natural) numbers expressed in either word or figure. Whewell thus extends the notion of an instrument at a time when scientific instruments such as air pumps, microscopes or telescopes are regarded mainly as means of doing physical experiments or conducting observations – they are deemed to be instruments suitable to test theories and to produce evidence or new phenomena, depending on the context and epistemic interests of the particular science in question.

When Whewell accepts *words and figures* as instruments; however, he imagines the act of inscription as an experimental practice. This has several consequences: Whewell's research instruments are not necessarily physical things in the sense of machines, laboratory devices or artificially enhanced human skills (e.g. optical tools) but also inscription systems. In addition, observations as well as experiments afford experiences that can be written down in numbers, which in turn can be expressed as words or as figures: numbers can be used for either classificatory or mathematical purposes, and both of these uses are equally valuable for the improvement of knowledge about natural things. The experimental character of the instrument *inscription system* becomes even more apparent when it is construed as a heuristic tool. In this sense, writing down *words and figures* certainly does improve the botanical and zoological classification and, what is more, it affords new systematic categories that eventually facilitate the prediction of unknown things or species.[22]

Some of Whewell's claims concerning the experimental importance of form serve as a reminder of Bacon's emphasis on the 'written experience' and of his development of the three types of tables in order to make similarities of instances visible, to display deficiencies and gaps, and ultimately, in the decisive step of an inductive inference that generates more general axioms, to reject certain instances. Whewell goes beyond this by designing far more sophisticated tables, notably arithmetic tables. More especially, however, he uses these tables in more than what one might call their *indicative form*: he applies tables containing words or figures as a heuristic tool and, thus, in their *subjunctive form*, as shown above.

Classificatory systems can also be viewed in light of the experimental mode. To find and construct conceptions in accordance with the method of classification involves an important constructive task, namely, determining an inclusive type that characterizes a natural group. Whewell emphasizes that it is less important to identify the boundaries or exclusive properties of such groups and more important to describe a central type that 'possesses in a marked degree all the leading characters of the class'.[23]

> A Natural Group is steadily fixed, though not precisely limited; it is given in position, though not circumscribed; it is determined, not by a boundary without, but by a central point within; – not by what it strictly excludes, but by what it eminently includes; by a Type; – not by a Definition.[24]

He continues in the same vein:

> The prevalence of Mathematics as an element of education has made us think Definition the philosophical mode of fixing the meaning of a word; if (Scientific) Natural History were introduced into education, men might become familiar with the fixation of the signification of words by *Types*; and this process agrees more nearly with the common processes by which words acquire their significations.[25]

Again, the prominence accorded to natural history draws attention to a general feature of scientific method. Here, once again, Whewell emphasizes the processual character of building scientific concepts. According to him, new concepts do not just capture what has already been discovered in the relevant facts; the invention of a concept (or term) is itself the discovery. The construction of scientific concepts thus goes far beyond collecting propositions and connecting them in logical operations or even reducing concepts to mathematical algorithms. Concepts, for Whewell, are not just hypothetical tools that might lead from given facts to the anticipation of a law. Instead a 'notion is *superinduced* upon the observed fact. In each inductive process, there is some general idea introduced, which is given, not by the phenomena, but by the mind'.[26] Accordingly, it is the invention of a conception that brings into being the very idea that a particular kind of regularity obtains in things and processes – these regularities are therefore not merely anticipated, be it a physical law or a botanical system. The mind is not viewed as a stock of limited 'Fundamental Ideas', as a kind of pre-fixed and bounded repertoire of forms of intuition and categories, but as essentially open. Conceptual schemes are created by the mind as cases emerge and stimulate attention and curiosity.

Whewell's Political Economy of Science

These considerations of likeness and the instruments of natural history lead to Whewell's *law of continuity*, a principle that relates to issues of quantity and qualitative resemblance. The aim of this principle is to avoid errors based on flaws in data collection: 'It is a test of truth, rather than an instrument of discovery'.[27] Thus the law of continuity may 'often be employed to correct inaccurate inductions, and to reject distinctions which have no real foundation in nature'.[28] It helps to expose partiality while simultaneously establishing and affirming the idea of *Natura non facit saltus*. Whewell explains that 'a quantity cannot pass from one amount to another by any change of conditions, without passing through all intermediate degrees of magnitude according to the intermediate conditions'.[29] This is a common depiction of a political economy of science committed above all to tracking and tracing entities – the cycle of an element or compound, the conservation of mass, the botanical or zoological system – as a check on the attainment of truth. This commitment likewise governs the economy of passions of the scientist, who:

> allows no natural yearning for the offspring of his own mind to draw him aside from the higher duty of loyalty to his sovereign, Truth: to her he not only gives his affections and his wishes, but strenuous labour and scrupulous minuteness of attention.[30]

To be sure, Whewell's law of continuity draws on the conservation laws of physics: it is about conservation and limited resources, about controlling and maintaining balance, and it demands adaption to the limits and constraints given by nature. The law of continuity instructs scientific housekeeping in terms of accounting for matter and energy and the transformation of entities.[31] However, as Whewell points out, in addition to this conservative method there is a method that approximates very closely to the law of continuity and that 'may be employed as positive means of obtaining new truths',[32] namely, the *Method of Gradation*.

'Distinctions of Opposites' or 'Differences of Degree'

Whewell suggests that the method of gradation can be regarded as the 'main business of science'[33] because it helps to distinguish those cases that have to be grouped together for their resemblance from those that have to be kept separate because they are distinct. It offers tools for selecting instances which agree and those which differ, thereby setting out 'important steps in the formation of science'.[34] It is this work of selection that determines the order of instances and therefore, ultimately, the configuration of facts. It is therefore fundamentally important to find evidence that allows one to confidently and reliably differentiate whether classes of things and properties 'are separated by distinctions of opposites, or by differences of degree'.[35] It is precisely here that experiments come into play. The method of gradation 'consists in taking intermediate stages of the properties in question, so as *to ascertain by experiment* whether, in the transition from one class to another, we have to leap over a manifest gap, or to follow a continuous road'.[36] In his method of learning through written experience, Bacon was aiming at something similar by devising inductive tables and thereby establishing classes of instances such as the *instantiae absentiae in proximo* and the *instantiae migrantes*.

In contrast to this, Whewell insists that this 'could hardly lead to scientific truth'[37] or real knowledge because Bacon takes 'the Natures and Forms of things and of their qualities for the primary subject of our researches'[38] – instead of trying to arrive at what Whewell calls the law of phenomena. As a good example of what these laws are and how to arrive at them by the method of gradation, he refers to the studies conducted by the electrical philosopher Faraday on the question of whether matter – be it sulphur, lac, ice, spermaceti or metal – is always either a conductor or a non-conductor and should thus be regarded as existing within an essential opposition. In the course of his experiments Faraday arrives at the conclusion that, even in metals, the electric current can be retarded. He asks (thereby arguing for a difference of degree) 'why this retardation should not be of the same kind as that in spermaceti, or in lac, or sulphur?'[39] This example from Faraday confirms not only that differences of degree need to be distinguished from differences of kind but also Whewell's emphasis on the need to do this by way of experiment. In a stepwise manner, science moves through experimentation, attentive observation, and the systematic collection of data towards a proper determination and description of phenomena. Rather than take the phenomena as given, Whewell's inductive method serves as a robust law of phenomena. And it is in this law of phenomena that Whewell's philosophy of science coincides with the hypothetico-deductive model of scientific reasoning.

Colligation: Seeing the Facts Instead of Summing Them Up

When Whewell distances himself from Bacon's method of induction, he is not just referring to Bacon's inductive tables but rather to the fact that Bacon does not go far enough in the procedure of generalization, settling much too soon for 'the Natures and Forms of things and of their qualities'. For Whewell, the definitions and principles that have been reached via the 'stair[case]' of inductive inference are a desirable 'final *result* of the reasoning, the ultimate effect of the proof', whereas in deductive reasoning:

> the general principles, the Definitions and Axioms, necessarily stand at the *beginning* of the demonstration ... The doctrine which is the *hypothesis* of the deductive reasoning is the *inference* of the inductive process. The special facts which are the basis of the inductive inference are the conclusion of the train of deduction. And in this manner the deduction establishes the induction. The principle which we gather from the facts is true, because the facts can be derived from it by rigorous demonstration. Induction moves upwards, and deduction downwards, on the same stair ... Deduction descends steadily and methodically, step by step: Induction mounts by a leap which is out of the reach of method. She bounds to the top of the stair at once; and then it is the business of Deduction, by trying each step in order, to establish the solidity of her companion's footing. Yet these must be processes of the same mind.[40]

Once again it becomes evident that in Whewell's philosophy the invention and subsequent corroboration of conceptions are central to the inductive process which, at the same time, involves the *deliberate imposition upon the data*. The antithetical principle of induction and deduction offers a means of first assembling and unifying phenomena and, in a second step, of proving the conditions of their connection. Induction always starts with this mental act: a colligation is a way of *seeing* the facts; it is 'never the mere *sum* of the facts', because it interprets the facts '*in a new light*'.[41] The introduction of novel concepts or propositions promotes a selective aggregation of facts (the process of colligation), which is not only a heuristic tool in the context of discovery but, due to the strong connection between deductive and inductive steps, provides us with a theory of confirmation.

This is certainly a more precise and therefore more powerful method for establishing and systematizing knowledge than Bacon was able to offer. Whewell successfully presents a method of induction that is necessarily related to a method of deduction by explicitly following certain principles of unification and simplification: theories and hypotheses can be tested on various levels, a proposition relating to certain data or facts is put forward that claims to be true, the colligation of a given body of data either works or fails. Conceptions that can be quantified are held to be more successful in providing good inductions than those that cannot. The more facts can be subsumed within a conception via colligation, the closer we come to a perfect table of induction for a science, that is, to a representation of all possible facts in the scheme – one that is perfectly true:

> The table of the progress of any science would thus resemble the map of a river, in which the waters from separate sources unite and make rivulets, which again meet with rivulets from other fountains, and thus go on forming by their junction trunks of a higher and higher order. The representation of the state of a science in this form would necessarily exhibit all the principal doctrines of the science; for each general truth contains the particular truths from which it was derived, and may be followed backwards till we

have these before us in their separate state. And the last and most advanced generalization would have, in such a scheme, its proper place and the evidence of its validity.[42]

According to Whewell, truth is expressed (and can only be expressed) by conceptions which give rise to his 'laws of phenomena', an expression which, in the
final analysis, proves to be coterminous with the laws of nature.[43]

> It must be tried by applying to the facts the conceptions which are derived from the
> idea, and not accepted till some of these succeed in giving the law of the phenomena.
> The rightness of the suggestion cannot be known other than by making the trial. If
> we can discover a *true law* by employing any conceptions, the idea from which these
> conceptions are derived is the *right* one; nor can there be any proof of its rightness
> so complete and satisfactory, as that we are by it led to a solid and permanent truth.[44]

In moving up the ladder of induction and deduction and climbing towards more
general and algorithmically enriched conceptions, the empirical data and facts are
progressively dematerialized, while the relation between theory and fact, between
thoughts and things, comes closer and closer to an ideal form. In the end, facts
are merely materially disembodied embodiments of hypotheses. However, this
increased formalization towards general laws and rules is countered by an explicit
realism which is also manifested in his 'unmistakable empirical emphasis'.[45] The
antithetical principle does not allow us to assign preference to any one direction
of explanatory power (from general to particulars or from sense data to laws) but
instead requires trust in this constant shift in direction. This is of utmost importance for Whewell because his theory of induction claims to be open towards
learning and an increase of knowledge; it therefore needs to include conjectural
processes.[46] This is why the construction of hypotheses,[47] the introduction of new
conceptions, and the subsequent colligation of facts prior to possessing evidence
in full are so important in the process of arriving at more generality.[48]

For Bacon, too, the method of induction is a heuristic tool, but it is *just* a
heuristic tool, based mainly on mechanisms of exclusion; in other words, it is
primarily 'eliminative induction'.[49] Bacon's interest in the 'negative cases' is
driven by his belief that it is possible to find the ultimate explanations if only
we can manage to weed out those factors that are not necessary to the production of an effect. Another important difference is that Bacon is committed to an
ontology of things. Bacon is interested in laws and causal structures, not merely
in order to better understand the material world but as a means to transforming
nature for human purposes: 'Lawlikeness is the crucial connection between our
knowledge of basic structure and our ability to transform nature'.[50]

Manufacturing Experiments: The Artful Work of Hands

Moreover, his emphasis on art and labour in the manipulation of material and in affording things also makes Bacon's theory of the basic structure of things attractive for contemporary philosophers of science concerned with experimentation, technoscience and emerging technologies as well as with the changing relations of science and society and of innovation and utility. By the same token, Bacon is frequently referred to by today's sociologists and political scientists, particularly in relation to concepts such as co-production and co-action.[51] And though Bacon's philosophy of science lacks the complexity and subtlety of Whewell's, his idea of a middle path involving matter and mind in equal measure in an ongoing process of inventing and systematizing things might be seen as one important building block underpinning Whewell's antithetically structured epistemology.

Despite his emphasis on the power of conceptions and his commitment to generalized ideal forms, Whewell never loses sight of the important role of manipulation in the process of testing conceptions through observation and experiment. The skills and proficiency of scientists determine their capacity to find and use those material forms by which 'nature shall be asked the question which we have in our minds'.[52] Whewell calls this 'a peculiar Art'. In order to illustrate the close interaction between mind and hand and between language and agency, Whewell refers to chemistry as an example:

> the chemist must learn to interpret the effects of mixture, heat, and other Chemical agencies, so as to see in them those facts which chemistry makes the basis of her doctrines. And in learning to interpret this language, he must also learn to call it forth; – to place bodies under the requisite conditions, by the apparatus of his own laboratory and the operations of his own fingers. To do this with readiness and precision, is, as we have said, an Art, both of the mind and of the hand.[53]

When emphasizing so zealously the importance of practice in the making of scientific knowledge, Whewell exhibits (again) his appreciation of Bacon's experimental philosophy. Again, however, he also distances himself from Bacon when pointing to facts rather than objects, as Bacon might have preferred. Whewell never misses an opportunity to highlight the hierarchical dependence of both the arts and engineering on science: only the latter is able to systematically study the properties of things and to provide the necessary theoretically based knowledge.

> [I]n the case of the improvements of the steam engine made by Watt, we have an admirable example how superior the method of improving Art by Science is, to the blind groupings of mere practical habit. Of this truth, the history of most of the useful arts in our time offers abundant proofs and illustrations. All improvements and applications of the forces and agencies which man employs for his purposes are now commonly made, not by blind trial, but with the clearest theoretical as well as practi-

cal insight which he can obtain, into the properties of the agents which he employs. In this way he has constructed, (using theory and calculation at every step of his construction,) steam engines, steam boats, screw-propellers, locomotive engines, railroads and bridges and structures of all kinds.[54]

In this passage, Whewell paints a picture of science as a child of enlightenment:[55] science helps us to 'see' and to purify our knowledge, be it of a theoretical or practical kind. Science allows us to go beyond the methods of 'blind trial' and 'blind groupings of mere practical habit' and to identify and describe the forces and agencies of matter. Since this is a science that is clearly directed towards an ideal of control and regulation and thus towards dispositional knowledge (*Verfü-gungswissen*), one might wonder whether this implies a tension with Whewell's constant expressions of commitment to truth.

Consilience of Inductions: Capstone or Irritant?

So far, colligation has been discussed as a way of explaining what has already been observed. In Whewell's view, however, truly effective colligation goes beyond this. In this view, colligation can be understood as an explanation based on reliable evidence, provided it successfully predicts phenomena not yet observed. In this case, the original idea of colligation is reaffirmed by a further accumulation of predicted observations that turn out to be correct. Whewell goes beyond this requirement, however, when he argues that the best test of any scientific explanation or theory is not prediction but *consilience*. This notion constitutes the 'leitmotif'[56] in the *History of the Inductive Sciences* (1837) and is thus of major importance to Whewell's logic of induction as a tool to describe the progressive nature of scientific growth and evolution. A consilience of inductions is said to take place when two chains of inductive inferences from what were originally different classes of phenomena prompt us to reach the same conclusion.[57] Modifying Whewell's above quoted metaphor, a consilience of inductions might consist in the realization that a heap of pebbles and a heap of pearls can each be threaded together to form a beautiful piece of jewellery. 'Consilience is in this sense', as Robert Butts puts it, 'a license for changing the semantics of earlier theories' – the original colligation of pearls now appearing as an instance of threading things together in a particular way.[58] It contributes to a reduction of concepts while simultaneously increasing substantive content: 'The Consiliences of our Inductions give rise to a constant Convergence of our Theory towards Simplicity and Unity'.[59]

According to Whewell, consilience leads to a type of theory that is extremely successful because it needs to employ only few predicates, albeit these few are highly demonstrative. Such theories possess greater scope and generality than others. They have achieved a certain unifying power, which Whewell takes to be equivalent to being simple or simply evident.[60] This is borne out by one of the most important examples in Whewell's argument regarding consilience, namely, his detailed study of the history of Newton's laws. The crucial step was to discover that satellite motion, planetary motion and falling bodies obey the same laws of motion. Newton realized that an inverse-square attractive force provides a causal means to describe various previously scattered phenomena: 'What Newton did, in effect, was to subsume these individual event kinds into a more general kind composed of sub-kinds that share a kind essence. Consilience of event or process kinds results in *causal unification*'.[61] The causal unification afforded by a consilience of inductions has the potential to be perceived as a scientific revolution that constitutes a research tradition or scientific discipline.

> We shall frequently have to notice the manner in which great discoveries thus stamp their impress upon the terms of a science; and, like great political revolutions, are recorded by the change of the current coin which has accompanied them.[62]

This is reminiscent of Lakatos's notion of *research programme*: a research tradition embodies basic theories that become either more complex and artificial (and that tend to display 'degenerating' complications) or more general and simplified, at the same time extending the theory to a wider domain of phenomena, in the manner of a progressive research programme.[63] The basic theories are constantly being reformulated by the scientific community, mainly by reconciling new phenomena with the theory. For Whewell, as in the case of a Lakatosian research programme, theories cannot simply be dismantled by an empirical argument but are rather weakened by the fact that the basic underlying theory has become too complex, unwieldy and logically untidy.

With regard to experiment and observation, the most pertinent question about consilience is how evidence can be provided, and this comes down to the question of what theory of confirmation is at work here.

> If we take one class of facts only, knowing the law which they follow, we may construct an hypothesis, or perhaps several, which may represent them: and as new circumstances are discovered, we may often adjust the hypothesis so as to correspond to these also. But when the hypothesis, of itself and without adjustment for the purpose, gives us the rule and reason of a class of facts not contemplated in its construction, we have a criterion of its reality, which has never yet been produced in favour of falsehood.[64]

Considering the success of a hypothesis as a criterion of its reality is another indication of Whewell's realism. This appeal to the realism of success also lurks in the background when Whewell explains his concept of consilience as the jumping together of inductions.

> Accordingly the cases in which inductions from classes of facts altogether different have thus *jumped together*, belong only to the best established theories which the his-

tory of science contains. And as I shall have occasion to refer to this peculiar feature in their evidence, I will take the liberty of describing it by a particular phrase; and will term it the *Consilience of Inductions*.[65]

Considering this point more closely and systematically, especially with regard to questions of observational and experimental evidence, it is useful to distinguish with Larry Laudan three types of consilience, which he sees as being 'three slightly different methodological gambits'.[66] The first one has already been mentioned. It is a consilience that simplifies and unifies theories and hypotheses without expanding on the empirical content. It is thus a type of consilience that provides us with formal and/or systemic features but not with empirical ones. The second type becomes salient in the quote cited above from the *Novum Organum Renovatum*: the adoption of the new hypothesis anticipates previously unknown natural phenomena, meaning that the explanatory content of the science is increased. As pointed out above, for Whewell it is this predictive capacity that is of utmost importance in justifying the significance of the inductive method for the progress of science. The third type of consilience concerns the 'severity of the various tests', as Popper called it. In other words, a theory appears to be all the more convincing the better it can integrate especially surprising effects that might otherwise have falsified it. This means (to continue in Popper's terms) that stronger theories are the ones that have been or that can be subjected to stronger tests. And yet the degree of corroboration provides no guarantee regarding the theory's future performance. The content of a theory, that is, the information embodied in corroborated and integrated facts, stands in an inverse relation to probability: the more content a theory has or the more consilient it is, the less probable it is – and low probability of a theory correlates to a high degree of falsifiability.[67]

Laudan believes that Popper and Whewell not only had quite similar ideas but that 'Popper's major "discovery" of the 1950s was a re-formulation of the problem of consilience ... [I]n his many discussions of the criteria for *severe* tests, [he] has required a "good" hypothesis to do precisely what Whewell expected it to do'.[68] Both of them insist that the measurable quality of a theory or hypothesis can be identified with the way it accommodates intriguing predictions, incorporates predicted phenomena – thereby increasing content – and thus makes it possible to explain phenomena of different kinds.[69]

One does not do justice, however, to Whewell's experimentalism and his inductive philosophy of invention and discovery if one considers the consilience of inductions merely in the context of justification and evidence. This becomes apparent from Barbara Stafford's criticism of E. O. Wilson's appropriation of the notion of consilience.[70] In his book *Consilience – the Unity of Knowledge*, the prominent biologist Edward O. Wilson refers to Whewell only in the first few pages when he speaks of 'a "jumping together" of knowledge by the linking of facts and fact-based theory across disciplines to create a common groundwork of explanation'.[71] Wilson goes on to use consilience mainly to denote knowledge transfer between disciplines and the development of 'shifting hybrid domains', and he finishes by identifying a 'consilience world view' in the following terms: 'all tangible phenomena, from the birth of stars to the workings of social institutions ... are ultimately reducible, however long and tortuous the sequences, to the laws of physics'.[72] This is surely not what Whewell had in mind when he visualized the progress of scientific knowledge using an alluvial metaphor, comparing it to a map of a river in which waters from different sources converge to form 'trunks of a higher and higher order'.[73] Where Wilson points downward to a fundamental bottom line, Whewell points upward to higher generalities. Laura D. Walls has summed up this misreading in the phrase: 'where Wilson is reductive, Whewell is additive'.[74]

Furthermore, Wilson also uses consilience as a method for mapping what is known and what are 'still unexplored domains of reality'.[75] In her book *Visual Analogy: Consciousness as the Art of Connecting* (2001), Barbara Stafford responds to this by pointing out that Wilson's notion of consilience is driven largely by a hierarchical reductivism and that he pursues an 'isomorphic program for intellectual annexation, whereby one branch of learning is subsumed under a more "fundamental" or powerful science'.[76] Stafford's criticism is motivated by her insistent concern to rehabilitate analogy as an epistemic tool and as a way of thinking that guides our attention towards making connections or 'joints'. Analogy finds coherences instead of constantly identifying differences and idealized dichotomies to the exclusion of in-betweens or third things. As we have seen in the discussion of resemblance and representation and of natural history as a science on a par with the mathematical and physical sciences, Stafford's quest for the construction of 'a more nuanced picture of resemblance and connectedness'[77] aligns very well with Whewell's conception of consilience.[78]

> We cannot read any of the inscriptions which nature presents to us, without interpreting them by means of some language which we ourselves are accustomed to speak; but we may make it our business to acquaint ourselves perfectly with the language which we thus employ, and to interpret it according to the rigorous rules of grammar and analogy.[79]

Embracing the Threshold of Belief

As we have seen from our discussion of Popper, Lakatos and Laudan, but also from the difference between Wilson and Stafford, 'considerations akin to consilience have found their way into almost every recent account of scientific method', but the 'basic problem to which Whewell addressed himself is still unresolved'.[80] For Laudan it is particularly noteworthy that 'the impressive array of formal tools of analysis'[81] currently available does not allow us to advance any further in questions of evidence and justification than Whewell was able to with his account of consilience. According to Whewell, every successful consilience – even after its most critical conceptual evaluation and the most severe testing of theory – involves a 'threshold of belief' which may be akin to the commitment to a research programme or a scientific paradigm. This 'threshold of belief' is what Laudan regards as the problem that Whewell left behind and that remains unresolved – it is as irreducible as is the interplay of invention and justification.

> Although in every inductive inference, an act of invention is requisite, the act soon slips out of notice. Although we bind together facts by superinducing upon them a new Conception, this Conception, once introduced and applied, is looked upon as inseparably connected with the facts, and necessarily implied in them. Having once had the phenomena bound together in their minds in virtue of the Conception, men can no longer easily restore them back to the detached and incoherent condition in which they were before they were thus combined. The pearls once strung, they seem to *form a chain by their nature* [emphasis added]. For instance, we usually represent to ourselves the Earth as *round*, the Earth and the Planets as *revolving* about the Sun, and as *drawn* to the Sun by a Central Force; we can hardly understand how it could cost the Greeks, and Copernicus, and Newton, so much pains and trouble to arrive at a view which to us is so familiar. These are no longer to us Conceptions caught hold of and kept hold of by a severe struggle; they are the simplest modes of conceiving the facts: they are really Facts. We are willing to *own* our obligation to those discoverers, but we hardly *feel* it: for in what other manner (we ask in our thoughts) could we represent the facts to ourselves.[82]

A Case Study: Stepping Outside – Victorian Tidal Studies

In early nineteenth-century Britain, tidal studies were in fashion, promoted not only by the scientific community but also by the British government and military. These studies were part of a broad movement of global geophysical research initiatives and, beginning in the middle of the century, of biological initiatives as well. They included research on tides, magnetism, meteorology and ocean currents, with interests soon expanding to include living creatures, in particular the systematics and distribution of planktonic and deep-sea organisms. All these projects 'expertly demonstrate the increasingly productive relationship between the British Admiralty and the scientific elite in Britain'.[83] The study of tides was not only of particular interest for practical reasons but was also a matter of prestige for the British nation, since the equivalent rival French academy had focused on terrestrial magnetism.

When Whewell's interest in tidal studies was triggered, the general law governing the tides had been already discovered by Newton, who had used tidal data from maritime journeys as supporting evidence for his theory of universal gravitation. However, it was not possible to deduce accurate data for specific places – e.g. for British harbours – from the general law, which gave only a rough idea as to when the tides actually occurred along the coasts. In collaboration with astronomer and mathematician John William Lubbock, Whewell started to collect data from long-term observations and immersed himself in the theoretical study of tides as a means of determining the correct mathematical variables for Newton's general tidal theory. Whewell and Lubbock went on to produce timetables for the most important British ports, London and Liverpool, and later extended their studies to other sites as well. This process turned out to be rather time-consuming, and so Whewell decided to try out a new theory: a calculation of the tides based on their progressive movement across the oceans. The basic underlying idea was that the known course of the tides along with accurate data for tidal movements and times at one port should enable researchers to extrapolate from this one port to the next one – and so on to include all ports in Britain, Europe and beyond. To arrive at such a theory of tidal progression, it should be sufficient to possess data based on short-term observations, though preferably from a large number of ports all over the world. In close collaboration with the admiralty and supported by a grant from the British Academy, Whewell subsequently organized what was probably the first global observation network. The data needed not only to be normalized using a standard method of measurement but also collected simultaneously in all measuring stations that took part in the programme.

Whewell successfully organized this measuring campaign, which included professional researchers and lay observers. It started in Great Britain, Ireland and France and later took place in other parts of the world with the help of the British hydrographic office and international contacts to foreign researchers who were studying hydrography. 'In June 1835, nine countries and many of their surveyors and explorers took measurements of the tides every fifteen minutes, day and night, for two weeks.'[84] In total, more than 650 tide stations contributed data.[85] These were subsequently collected, reduced and represented in tables, graphs and, finally, in maps (see Figure 3.1).

Tides of the Arctic Ocean.

The main tide-wave which we have been following, after reaching the Orkneys, will, I conceive, move forwards in the sea of which the shores of Norway and Siberia form one side, and those of Greenland and America the other. It will here meet in succession with the islands of Iceland and Spitzbergen*; it will pass the pole of the earth, and will finally end its course on the shores in the neighbourhood of Behring's Straits. Perhaps it may propagate its influence through the Straits, and modify the tides of the North Pacific.

But a branch tide is sent off from this main tide into the German Ocean: this, entering between the Orkneys and the coast of Norway, brings the tide to the east coast of England, the coast of Holland, Denmark, and Germany. We shall now trace this tide.

* I add the following statements, which refer to the further course of the tide.

Bergen, lat. 60° 24′	1ʰ 30ᵐ	NORIE.
Drontheim, lat. 63° 26′................	2 15	——
Hammerfest, lat. 70° 40′	1 10	——
North Cape, lat. 71° 10′	3 44	——
Sweetnose (Lapland), lat. 68° 10′	8 30	——
Isle Kilduin (Lapland), lat. 69° 10′......	7 30	LALANDE, p. 340.
Archangel...........................	6 0	——, pp. 272 & 340.
Patrix Fiord (Iceland), lat. 65° 36′......	6 0	——, p. 340.
Hakluyt's Head (Spitzbergen)..........	1 30	beginning of tide. PHIPPS, pp. 44 & 67.
Magdalen Bay	1 30	PHIPPS, p. 30.
Moffen Isle, 25th July 1773, low water at	′1 0	——, p. 50.

Figure 3.1: Data for the tides of the Arctic Ocean, in W. Whewell, 'Essay towards a First Approximation to a Map of Cotidal Lines', *Philosophical Transactions of the Royal Society of London*, 123 (1833), pp. 147–236, on p. 184. Reproduced courtesy of the Library of the University of Basel, Switzerland. By 1833 Whewell had published a paper that presented the data generated in the first measurement campaigns. The paper is accompanied by numerous tables and two maps, one of them representing 'the greater part of the world', the other one the British Isles. At the same time the paper served to announce the need for further research. Whewell was especially in favour of having more simultaneous tide observations conducted at different places on the same stretch of coastline: 'These would enable us better than any other observations to determine the motion of the tide-wave along the coast' (p. 236).

Whewell produced an isotidal map based on a type of representation previously developed by Humboldt (1908) to map isothermal conditions in the Andes during the latter's expedition in South America. These maps made it possible to view at a glance how the tides progressed through the Atlantic and the Pacific and finally onto the shores of the European empires. The tidal venture became the prototype for a 'global science' that was also applied to the study of terrestrial magnetism, ocean currents and similar phenomena (see Figure 3.2 and Figure 3.3).

Whewell succeeded in stringing together the general law of gravitation, the theory of equilibrium (of fluids), a 'special empirical law' of the tides and, finally, the tidal observations that were represented in another type of table and compiled by Whewell's co-worker Lubbock. Despite this success, however, Whewell was not at all sure whether the 'special empirical law' of the tides was a theoretically convincing theory or was perhaps even a false one. He concludes his account by asserting that there are important disagreements between the theoretical laws and the facts (although there was no doubt that the theoretical formula he developed was fairly powerful). 'A little while ago the theory was in advance of observation; at present observation is in advance of theory'.[86] In formulations like these, the author of a philosophy of induction is already beginning to raise his voice.

Tidal Practices

In Whewell's view, the main problem of the 'special empirical law of the tides' is that the equilibrium theory and the Laplacian theory were not adequate to start with. Instead, a theory of the motion of fluids was needed, he believed, which unfortunately was not yet available because of the 'extremely imperfect state of the mathematical science of hydrodynamics'.[87] From this starting point, Whewell develops a series of formulas to determine and predict the height and time of high water at Liverpool, and was thereby able to model a data set which included the data set gathered from about ninety observations over nineteen years. On the basis of Laplace's assumed inadequate theory concerning the motion of the celestial bodies and lunar and solar forces, Whewell makes a series of calculations to eliminate the effect of the lunar parallax (the effect of changes in lunar declination) and thereby determines the solar inequality of the heights and times and the diurnal inequality of the tides at Liverpool (see Figure 3.4).

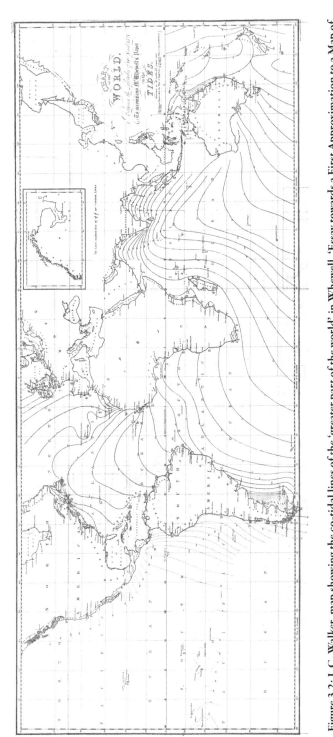

Figure 3.2: J. C. Walker, map showing the co-tidal lines of the 'greater part of the world'; in Whewell, 'Essay towards a First Approximation to a Map of Cotidal Lines', p. 147, plate V. Reproduced courtesy of the Library of the University of Basel, Switzerland. This map, drawn and engraved by Walker, is part of a first series of maps produced in the course of Whewell's studies of the tides. What shows up as a palimpsest is the same map but horizontally mirrored and inverted. Some of the printed numbers and lines end only beyond the map frame (at the tip of South America). Thereby the map becomes a bit of a sketchy character, albeit being at the same time incredibly detailed. However, Whewell noted: 'I shall be neither surprised nor mortified if the lines which I have drawn shall turn out to be in many instances widely erroneous: I offer them only as the simplest mode which I can now discover of grouping the facts which we possess' (p. 235).

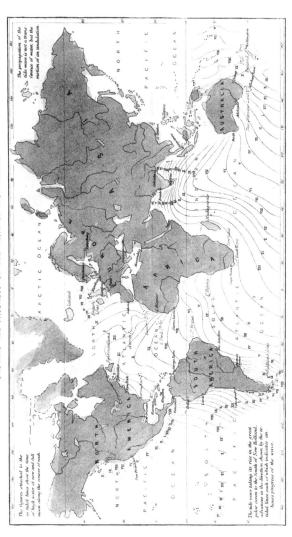

Figure 3.3: John Emslie, 'Tidal Chart of the World: Showing the Progress of the Wave of High Water' (1850). Reproduced courtesy of the Science & Society Picture Library, London. Emslie, who drew and engraved this tidal map, used Whewell's map of co-tidal lines to demonstrate the dynamics of the tide through the ocean and some major ports. Each of the co-tidal lines indicates the direction of advancement and the hourly progress of the wave. The numbers attached to the lines give the time of high water at new and full moon. The map was printed on one page together with a botanical map showing 'the distribution of plants and their cultivation over the world'. This page in turn was part of a portfolio collection of forty-four scientific teaching diagrams, all of them drawn and engraved by Emslie. The collection was produced for educational purposes and covered the fields of geology, geography, astronomy and natural philosophy. It is taken from a series published between 1850 and 1860 (probably written by James Reynolds) in response to the popular demand for information on the developments taking place in science and engineering as a result of the Industrial Revolution.

Table VII. (a.)

Mean of each column subtracted from the column "*Interval*" of times.

H. P.......	54'.	55'.	56'.	57'.	58'.	59'.	60'.	61'.	Mean.
Mean Interval }	h m 11 12·7	h m 11 11·5	h m 11 8·3	h m 11 6·5	h m 11 3·7	h m 11 0·3	h m 10 58·5	h m 10 54·5	h m 11 6
☽ 's Transit.	Remainder.	Remainder.	Remainder.	Remainder.	Remainder.	Remainder.	Remainder.	Remainder.	Remainder.
h m									
0 30	+13·7	+11·6	+ 8·0	+11·6	+12·2	+12·9	+13·3	+11·8	+12·2
1 30	− 5·2	− 6·8	− 6·5	− 3·9	− 5·3	− 2·8	− 2·5	− 2·8	− 4·6
2 30	−23·2	−22·0	−22·0	−20·8	−18·6	−17·5	−16·6	−17·1	−20·0
3 30	−41·4	−36·7	−34·7	−33·0	−32·1	−29·3	−30·4	..	−33·9
4 30	−49·0	−47·8	−44·0	−41·8	−40·4	−38·4	−38·7	..	−42·8
5 30	−47·7	−45·7	−43·6	−43·2	−39·4	−38·5	−37·5	..	−43·2
6 30	−27·2	−26·4	−24·5	−25·7	−25·6	−24·8	−21·5	..	−25·0
7 30	+14·3	+13·4	+11·9	+ 9·2	+ 6·5	+ 1·6	+11	..	+ 9·6
8 30	+44·8	+41·8	+40·1	+37·4	+34·1	+31·1	+20·2	..	+36·6
9 30	+50·9	+49·3	+47·9	+45·0	+44·1	+41·6	+39·8	+39	+45·6
10 30	+42·5	+41·8	+41·3	+40·1	+38·6	+38·0	+36·1	+35·8	+39·8
11 30	+28·5	+28·1	+26·6	+25·7	+23·6	+25·7	+24·4	+25·4	+26·1
Max. Diff.	99·9	95·1	91·9	88·2	84·5	80·1	78·5		88·8

On comparing the mean numbers in the last column with the theoretical formula

$$\tan 2\,(\theta' - \lambda') = -\,\frac{c \sin 2\,(\phi - \alpha)}{1 + c \cos 2\,(\phi - \alpha)},$$

it appears that they may be very accurately represented by making $\lambda' = 11^{\mathrm{h}}\,6^{\mathrm{m}}$, $\alpha = 1^{\mathrm{h}}\,15^{\mathrm{m}}$, $c = \sin 1^{\mathrm{h}}\,29^{\mathrm{m}}$. The agreement of this formula with observation is as follows:

Moon's Transit.	Formula.	Obs.	Diff.
h m	m s	m s	m s
0 30	+12 16	+12 12	− 0 4
1 30	− 4 7	− 4 36	− 0 29
2 30	−20 6	−20 0	+ 0 6
3 30	−34 0	−33 54	+ 0 6
4 30	−43 6	−42 48	+ 0 18
5 30	−42 40	−43 12	− 0 32
6 30	−25 8	−25 0	+ 0 8
7 30	+ 9 2	+ 9 6	+ 0 4
8 30	+36 28	+36 36	+ 0 8
9 30	+44 30	45 36	+ 1 6
10 30	+39 40	39 48	+ 0 8
11 30	+27 36	26 6	− 1 30

This accordance is complete, the difference amounting in only two cases to 1ᵐ.

Figure 3.4: Tables and algorithms showing tidal practices, in W. Whewell, 'Researches on the Tides. Fourth Series. On the Empirical Laws of the Tides in the Port of Liverpool', *Philosophical Transactions of the Royal Society of London*, 126 (1836), pp. 1–15, on p. 8. Reproduced courtesy of the Library of the University of Basel, Switzerland. These tables and algorithms give an impression of the path of increasing abstraction of the observed instances at shores worldwide. They show a comparison between observed numbers and the results obtained from a formula designed to describe the semi-monthly inequality of the times of high water for the port of Liverpool. Whewell considers the resulting agreement as sufficiently close to be valid, 'the difference amounting in only two cases to 1m' (p. 8).

Despite the constraints of an inadequate theoretical basis and his explicit statement that the special empirical law is 'not the true theory', Whewell insists that 'the tide at any place occurs in the same way as if the ocean imitated the form of equilibrium corresponding to a certain antecedent time', and that because of this the so-called *Equilibrium Theory* expresses with 'very remarkable exactness most of the circumstances of my result'.[88]

Clearly Whewell made an active contribution to the fashionable study of tides by generating original research results and by producing a novel research design, namely, isochronic data collection on a global scale suitable for determining and mapping the advancement of the waves in the oceans. His work in tidal research was met with general approval, and in 1837 Whewell was awarded a gold medal by the Royal Society of London. In the 1820s, though, prior to his prizewinning tidal studies, Whewell had already delved into questions relating to a mathematically founded crystallography, and his collaborative studies with German mineralogist Friedrich Mohs on a natural system of classification for minerals resulted in a monograph on mineralogy and several individual papers.[89]

All this shows that Whewell had first-hand knowledge in several areas of research when he started his work on the history and philosophy of the sciences. Moreover – and perhaps more importantly – his tidal studies indicate the deep roots in scientific practice of Whewell's antithetical epistemology and the idea of a balance of mind and hand. Convinced that 'at present observation is in advance of theory', he gets involved in an activity that might be conceptualized as a field experiment, in which data are generated on a global scale – or at least in the world as seen by the British Empire. Whewell engages in the design of a system of places and times that finally allows him to test hypotheses and to develop a theory that in the end acquires the respectable label of a 'special empirical law of the tides'.

Whewell's ideas about what he called the inductive sciences were by no means acquired through historical studies alone: he was skilled in doing empirical research. 'Knowledge of both current scientific practice and the history of science were important to Whewell in developing his philosophy of science'.[90] Whewell's scientific skills were also exhibited in his interactions with scientists in very different fields, such as physicist Michael Faraday, glaciologist James David Forbes, astronomer and mathematician John Herschel and many others.

His recognition by the scientific community certainly helped to further his philosophical programme regarding the crucial importance of conceptions in empirical research. The negotiation between Whewell and Faraday on the electrical field, for example, culminated in the coinage of the concept 'polarity'.[91] However, his prolific creative capacity to find the 'right' wording and thus 'to make language "a perfect daguerrotype" of nature'[92] also drew critical comments and earned him the epithet 'verbarian Attorney General': 'Whewell defined those who were allowed to make new scientific terms and also defined the terms they could use'.[93]

Experiment and Observation, Considered Together

As noted above, Whewell did not place much emphasis on the distinction between experiment and observation as a means of systematically generating scientific facts. Moreover, he did not attach any judgement to one or the other. As a consequence, it was not particularly important to him to draw a dividing line between scientific fields based on their use of predominantly experimental or observational methods and tools. For Whewell, good science is first and foremost simply a matter of collecting as many data as possible, doing so as precisely as possible, constructing a workable mathematical model and thus finding out as much as possible about the real world. It follows, then, that it does not much concern him whether the data are generated by experiment or by observation.

In this respect Whewell is, for once, in agreement with John Stuart Mill, whom he otherwise attacks for his conception of a system of inductive rules.[94] Mill argues that we either find a situation in nature that corresponds to what we are looking for or we set up this situation by doing experiments: 'The value of the example depends on what it is in itself and not on how it was achieved'.[95] Whether or not we intervene in nature is not important: it is the representation of nature that ultimately counts; there is no *logical* distinction between data obtained by mere observation and by doing experiments. For Whewell, the choice between observation and experiment is more a question of finding the appropriate form to describe and to present the empirical data in a methodologically correct manner:

> The Education of the Senses is also greatly promoted by the practical pursuit of any science of experiment and observation, as chemistry or astronomy. The methods of manipulating [in chemistry] ... and the methods of measuring extremely minute portions of space and time [astronomy] ... are among the best modes of educating the senses for purposes of scientific observation.[96]

All this suggests that developing a systematic means of observation is the most important issue for Whewell, and not whether this observation takes place in the context of an experimental set-up in the laboratory or 'outside in the field', as in the case of the tidal studies. This impression is further confirmed when he notes:

> The framing of hypotheses is, for the inquirer after truth, not the end, but the beginning of his work. Each of his systems is invented, not that he may admire it and follow it into all its consistent consequences, but that he may make it the occasion of a course of active experiment and observation.[97]

Two things are of interest here: first, Whewell points out that the work of the hand – *active* experiment and observation – is the necessary counterpart to conceptual work; it is not just a tool used to justify or reject a hypothesis. It is able to challenge what he calls 'the system', one that serves to stimulate further inquiry, just as the aphorism with its subjunctive form allows it to operate as a heuristic tool.

Second, Whewell considers both observation and experiment to be a positive and deliberate act. Both are perceived as a critical means of proving and improving a conceptual system, with the experiment neither being elevated to the status of indispensability (in the role of a crucial experiment) nor being used merely as a tool for justifying a theory. Thus Whewell always keeps in mind his dialectical conception of a fundamental antithesis in philosophy according to which we always have to *recognize the existence both of thoughts and of things* when we are doing empirical research. However, this activity of observing things and experimenting with them also implies that there are practices and instruments involved that help scientists to bring forth the facts relevant to the particular scientific community of which they are members. A chemist, for instance, needs to be so

> familiar with his science, that he has the power of observing. He must learn to interpret the effects of mixture, heat, and other Chemical agencies, so as to see in them those facts which chemistry makes the basis of her doctrines.[98]

Only then is the chemist able to successfully 'place bodies under the requisite conditions by the apparatus of his own laboratory and the operations of his own fingers'.[99] Consequently, a Whewellian scientist is committed to an idea of science that involves both the mind and the hand. One might argue that this conception anticipates the turn towards experiment and practice in the philosophy of science in the latter years of the twentieth century, as represented, for instance, by statements such as this: 'New knowledge and explanations emerge not from ideas and theories alone but also from experimental design and instrumental intervention'.[100]

It is a bit of a stretch, perhaps, to view Whewell as a forerunner of the so-called practical turn. And yet his inductive philosophy does not fit squarely either into the scheme of philosophical approaches common in the second half of the nineteenth century and well into the twentieth century. It is the precision and openness of his thinking that makes his philosophy a rich source for a number of interesting perspectives – such as the balance between hand and mind, the importance of manipulation, the passion for the work of concepts as instruments of science and, last but not least, his antithetical epistemology.

> For although man may in this way please himself, and admire the creations of his own brain, he can never, by this course, hit upon the real scheme of nature. With his ideas unfolded by education, sharpened by controversy, rectified by metaphysics, he may *understand* the natural world, but he cannot *invent* it. At every step, he must try the value of the advances he has made in thought, by applying his thoughts to things. The Explication of Conceptions must be carried on with a perpetual reference to the Colligation of Facts.[101]

These last two sentences in particular read like an anticipation of what John Norton claimed to be a central issue in his material theory of induction, which 'traces the justification of an induction back through chains of licensing facts whose justification is part of the regular practice of science'.[102]

When does an Experiment Begin: Experiment and Observation Distinguished

Whewell does not make a qualitative distinction between data obtained in an experimental setting and those obtained by mere observation. In both cases it is necessary to choose carefully our perspective, our instruments and the material from which we seek to elicit the empirical data. Given Whewell's apparently blurred use of the concepts of experiment and observation, one might ask whether his tidal studies should be described as an experiment on a global scale or whether he was simply carrying out a very complex observation in a highly organized and mathematically sophisticated manner. In either case it seems clear that the conditions for both good 'colligation' and 'consilience' have been fulfilled: the (measurable) quality of a theory or hypothesis consists in the fact that it allows for intriguing predictions; it incorporates predicted phenomena (and thus increases content); and it provides explanations for different kinds of phenomena. Thus one might ask – to invert the title of Galison's book on experimentation in physics – when does an experiment *begin*?[103]

Obviously, the operations described above bear little or no relation to the description of experimental laboratory conditions that are commonly found in the literature. Likewise, it is quite clear that the procedure Whewell advocated is much more than what one might properly describe as a pure thought experiment because it involves empirical data in model building. His approach represents another type of experimentation, namely, one that can be thought of as a *field experiment*.

This notion of the field experiment needs to be understood in contrast to the experiment in its more common usage, that of instrumental manipulation and bench activity in the laboratory sciences: 'experiments consist mainly of instruments and tools standing or lying on the table of a scientist'.[104] The notion of the field experiment relies instead on a number of minimal conditions that must be met for an experimental intervention to occur. These are based on agreement among the broader scientific community or within a specific group of scientists and concern the following three questions: what kinds of sense data are to be collected (variable over time and space), how can these measurements be accomplished (destruction by intervention, training of different observers, etc.), and what boundaries are relevant and how should they be conceptualized (what or who is part of the system and what is not)?

On this basis, there is good reason to designate Whewell's case study on tidal characteristics as a field experiment: the data are collected in several places at the same time and by different observers, yet all the measurements are taken in exactly the same way, and all these data are subsequently tabulated and processed by a set of formulas, generating facts that did not exist before – as is the case with the isotidal map and the hypothesized equilibrium conditions that it represents.

Whewell's inductive philosophy – or, as he called it more precisely, his fundamental antithesis of philosophy – provides a good foundation for investigation and may even open up a line of inquiry capable of leading towards a proper description of the 'field ideal' of research in opposition to the 'lab ideal'. It provides a perfect foil for a more sustained juxtaposition of laboratory and field experiment. Bacon's and Whewell's extended conceptions of experimentation can now be rediscovered in rather more contemporary developments in the philosophy of science.

4 THE EXPERIMENT IN RECENT PHILOSOPHY OF SCIENCE

In the twentieth century, experimentation became a focus of interest for philosophers of science from only the 1980s onwards, when they began to pay attention to the role of scientific practices in knowledge building rather than dealing solely with theories and purely logical operations.[1] Ian Hacking's *Representing and Intervening* (1983) is regarded as the first and most influential book that highlights the significance of experiments: 'Experimentation has a life of its own'.[2] During a conference at the MPI Berlin in June 2005 on 'The Shape of Experiment', Hacking himself referred to the fact that many voices came 'to the fore in the 1980s, urging that we think about experiment as intensely as we had been thinking about theorizing. Mine only happened to be an early contribution'.[3] However, in chapter 9 of his above mentioned book, entitled '*Experiment*', Hacking starts with a bold statement about the blind spot afflicting philosophers of science who 'constantly discuss theories and representations of reality, but say almost nothing about experiment, technology, or the use of knowledge to alter the world'.[4] A few pages later, he continues:

> What is scientific method? Is it the experimental method? The question is wrongly posed. Why should there be *the* method of science? ... We should not expect something as motley as the growth of knowledge to be strapped to one methodology.[5]

Since then it has become commonly accepted that there is more to experimentation in the lab than just the testing of theories, and the question arose as to what comes first – theory or experiment. Another important problem regarding the relationship between theory and experiment arises when we accept that not all the sciences handle this issue in the same manner: '[T]he relationships between theory and experiment differ at different stages of development, nor do all the natural sciences go through the same cycles'.[6] There is now notably less agreement around the question of how experiments address theories and even less about whether or not theory always plays a role in experimental practice.

This is not the place to join in the chorus lamenting the disregard shown by philosophers of science towards scientific practices and instruments and indeed laboratory technical skills in general up until quite late in the twentieth century. Nonetheless, it is a finding that might appear all the more curious given that, in the nineteenth century, a few philosophers (and philosopher-scientists) had already been engaged in theorizing the experiment and the balance between hand and mind, as discussed in the previous chapter. Above all, both logical positivists and empirical inductivists – representatives of the two most influential philosophical schools in the first half of the twentieth century – attached great importance to experience but were obviously less concerned about how this experience was literally manufactured: *manu factus, fait à la main, made by hand* was not in the focus of those philosophers.

In recent decades this remarkable tension has been elaborated extensively and a weighty corpus of literature generated.[7] In the following, I will flesh out the most salient arguments that have been put forward and have given rise to the new research fields in philosophy of science known as experimentalism,[8] philosophy of scientific experimentation[9] and (albeit with a different emphasis again) experimental philosophy. Kwame Anthony Appiah noted in his presidential address to the Eastern Division of the American Philosophical Association in 2007: 'The recent return to these shores of the epithet "experimental philosophy" is – as one tendency in our profession might put it – a return of the repressed'.[10]

The complexity of the debate, after about two decades of academic activity, is impressive. The topics listed below are regarded as the most pertinent ones, not least with regard to the main focus of this work, namely, field experimentation and ecology (or the environmental sciences):

1. the role of theory in experimentation (in the laboratory);
2. the material realization of experiments (including the role of instruments);
3. modelling and experimentation (scaling issues, computer simulations and real-world experiments);
4. the role of the social or cultural in experimental design.

In the following, the topics 'role of theory in experimentation' and 'material realization of experiments' will be discussed in two fairly condensed sections. In doing so, no pretence is made to cover every facet of this multifaceted debate. Instead, the sections are intended as preparation for a discussion of the latter two topics, which are key to my overall thesis concerning the (de)liberation of the experiment. To this discussion the remaining chapters of Part II will be dedicated.

The Role of Theory in Experimentation (in the Lab)

Intense discussions over the question of how experiments relate to theories have sharpened existing antithetical positions in the debate over what comes first, theory or experiment. On the one hand there is the traditional and heavily normative position that experiment without theory is somewhat worthless and unnecessary, as no theory can be deduced from the observed facts. This is countered on the other hand by the statement that theory must not necessarily and at all times play a role in experimental practice, especially where it is exploratory. Another point of critique was directed at the linguistic fixation according to which only the representation of experience counts in science and that all scientific arguments therefore begin with observational statements – independently of whether one subscribes to an inductivist or deductivist methodology.[11]

The 'standard view' among philosophers of the role of experiments in science basically comes down to the experiment having two tasks: to answer the questions that theories and hypotheses address to the world; and secondly, to select the theory or hypothesis that provides the best answers in the sense of making the most accurate predictions. Accordingly, Popper was adamant that it is the theorist who must formulate research questions as precisely as possible before the experimenter sets to work. New knowledge can be achieved only by a scientific development that is driven by theory and not by experiment, by ideas and or by observation. Lakatos and Kuhn, two of Popper's most famous adversaries, concurred with him in this respect. Lakatos holds that a single empirical finding, or a single fact, cannot disrupt a theory (and thus cannot function as a crucial experiment) because the research programme is able to compensate for such irregularities by constructing ad-hoc hypotheses. And Kuhn, despite being a critic of logical empiricism, also subscribes to the tangential role of experiment. He, too, was sceptical about the role of the 'Baconian movement' in the formation of the classical sciences, though he granted its qualitative novelty: 'how did its existence affect the development of science? The contributions of Baconianism to the classical transformations of the sciences were very small. Some experiments did play a role, but they all have deep roots in the older tradition'.[12] In stating this, he concurs with his teacher Alexandre Koyré, who emphasized the importance of theory and of a mathematically oriented science over and against technology and experimentally driven science. Kuhn follows this scheme in proposing to group all scientific activities into two clusters, the classical and the Baconian.

> Crudely put, the classical sciences were grouped together as 'mathematics'; the Baconian were generally viewed as 'experimental philosophy' or, in France, as *physique expérimentale*; chemistry, with its continuing ties to pharmacy, medicine, and the various crafts, was in part a member of the latter group, and in part a *congeries* of more classical specialties.[13]

Rather reluctantly, Kuhn also acknowledges, however, that German successes in the transformation of physics 'must be due' to the accelerated growth and consequent plasticity and openness of educational institutions 'when men like Neumann, Weber, Helmholtz, and Kirchhoff were creating a new discipline in which both experimentalists and mathematical theories would be associated as practitioners of physics'.[14] This seems to suggest that a collaborative learning process in an interdisciplinary context generates a new scientific field and, in particular, that the association of experimenters and theorists is achieved through collaborative development and by rehearsing a particular scientific practice. Such a scientific practice is one that aims at integrating experimentation with things and ideas in equal measure, one which entails an emphasis on inductive procedures in knowledge building and does not prevent the application of deductive methods. All of this certainly held for Hermann von Helmholtz. Being at once a physician, physiologist and physicist, he acknowledged that the experiment plays an important role, a view reflected in his scientific practice as well as in his epistemological explorations:

> [I]n the experiment, the chain of causes runs through our consciousness. We can only know something about the causes once we have conducted the experiments. Only through inner contemplation can we gain an idea of what we have, through an act of will expressed in these actions, successfully caused to occur.[15]

This position is not as uncommon or isolated as it may appear given the dominant narrative in philosophy of science. The latter, however, remains resistant to efforts at revision and has yet to be substantiated with further historical case studies that challenge the experimentalism perspective.

A major impediment to the acknowledgement of a constitutive, originary role to experimentation in theory building resides in the famous distinction between the context of justification and the context of discovery, which was introduced mainly by Hans Reichenbach and was taken up by Karl Popper, but can be traced back to Gottlieb Frege. It is easy to see that the methodological primacy accorded to theories is only the flipside of the philosophical preoccupation with the context of justification, considered to be the principal concern specifically of the exact sciences as opposed to the so-called inexact sciences and even more to engineering or so-called pseudo-sciences. The context of justification is regarded as a domain of methodology whose proper task is to evaluate theories that have been constructed already. As for the context of discovery, it is taken to belong to the psychological sphere of creativity and invention. Here 'the process of discovery allegedly rests on what Einstein called "empathy with nature"'.[16]

To counter this conceptual obstacle, it was suggested that experiments may function simultaneously as a means of testing a hypothesis *and* as instruments of discovery. This is the case, for example, 'when an investigator is interested in a theory involving some free parameters, and he performs experiments simply to determine the values of these parameters'.[17] It is also the case when a notion of data-driven science is introduced to complement theory-driven science. The latter is familiar: data obtained by experiment or observation are used to test a pre-existing theory. In contrast, data-driven science advances a process of discovery: a body of data is taken as a starting point, and scientists search for a theory to explain these data. This distinction was proposed by social scientist Herbert A. Simon, who established a direct relationship between the two with his computer programme 'BACON'. In a simulation of the process of discovery, this 'data-driven inductive machine' is capable of 'performing factorial experiments' on numerical and nominal data.[18]

These are just two suggestions for adjusting theoretical concepts to more adequately reflect scientific practice and to better understand the role of the experiment in scientific theory building. They represent approaches that may be subsumed under a twofold conception of experimentation. Typically, experimentation is still committed to the boundary work of retaining the separation between the context of discovery and the context of justification. However, once the experiment is accepted as having relevance in both areas – in testing, modifying and verifying theories and in playing a heuristic role – 'the attendant antithesis of logic and psychology does not remain untouched'.[19] Now, the psychological sphere of discovery is no longer the sphere where randomness and arbitrariness reign and where knowledge arises from a sudden yet inexplicable stroke of genius. Instead, the 'psychology of invention' is replaced by a 'logic of discovery' and follows a 'rational heuristics'.[20] This has also been described as a mode of experimentation, in which 'investigative operations'[21] are accomplished and ultimately become interwoven in a network of self-instructing epistemic practices.

Once this step is taken, the seemingly rigid separation of a context of justification from a context of discovery becomes open to challenge. Accordingly, the question of what comes first – experiment or theory – proves to be a rather pointless chicken-and-egg question. A comprehensive conceptual framework has to take into account the intimate connectedness between theory, model and experiment in scientific practice. A number of conceptualizations have since been put forward, not only in the context of history and philosophy of science (HPS) but also by science and technology studies (STS). To name just a few, these conceptualizations include the experimental system[22] and the trading zone[23] or, in the philosophy of science, the apparatus/world complex[24] and model experiments.[25] Since all these conceptualizations are closely linked also to questions concerning the role of instruments and materials, they will be discussed in the following section.

The Material Realization of Experiments

We have seen that the philosophical debates of the last two decades pushed the issue of experimentation beyond simply questions of truth value and justification and brought to light other aspects of the experiment such as the genesis of theories and the dynamics of interrelation between instruments, machines or apparatuses and scientific objects. Accordingly, one important aspect of the debate focused on what is implied philosophically when novel knowledge and explanations derive not only from ideas and theories but also from experimental design and instrumental intervention. Furthermore, it was pointed out that the heuristic role of the experiment is not based on a mere trial-and-error method in the sense of a non-methodological, piecemeal procedure. Instead, 'explorative experimentation' is accomplished in a systematic manner.[26] In the following, four conceptions of experimental investigation are presented in brief, each of them involving a different emphasis on the role of materials (or organisms), the relationship between technology and science, and the power granted to hands and eyes.

Experimentation Creates Reality

Michael Heidelberger points out that we can talk about the theory-ladenness of experiments without necessarily maintaining that every experiment is theory laden. On the contrary, he insists, one important and interesting feature of experimentation is that the manipulation of machines involves creativity. He proposes that we conceptualize the instrument as an embodiment of paradigmatic power, which makes it possible to attribute the generative function of the experiment mainly to the use of instruments. Heidelberger clearly has the world of physics in mind when he suggests a distinction between three types of manipulating instruments that make for productive, representative and constructive experimentation.[27] He suggests that constructing an electrical energy source requires a productive experiment, whereas operating an already existing source that functions constantly permits constructive experimentation. Both types of experiment are conducted in order to find out more about the conditions that reliably support the production of the physical phenomenon. It holds for both types that 'instrument usage enjoys an unfettered existence with regard to theory'.[28] Even the third, representational type can develop a heuristic dimension when the content of a concept is expanded. Therefore, each of the three types of experimenting can operate independently of theory, extending beyond the context of justification. In all three cases, experiments emerge from a background of hypothesis and expectation, and yet (as pointed out by Kuhn) experimentation ultimately gives rise to unexpected novelty. 'Beyond the mere testing of theories, the experiment also serves to create reality'.[29] Thus experimentation without theory does happen in scientific practice.[30]

Material Raises Questions

Hans-Jörg Rheinberger took up the scientific concept of experimental system, widely used in the biological sciences, to elaborate his ideas about the dynamics of experimental research on organisms (or parts of them). He starts from the assumption that experimental systems have a considerable influence on the direction the experimental inquiry will eventually take. This implies that theories are of rather minor importance or at least have to be closely contextualized within the scientific practice in the laboratory. The direct result of these activities is not statements and theories but objects. Biological research begins with 'the choice of a system rather than with the choice of a theoretical framework'.[31] The choice of a system implies a decision about both the scientific object and, similarly, the technological objects that allow the experimenter to manipulate and represent this scientific object, thereby limiting it at the same time. Indeed, these two types of object are so closely correlated that there is not only an interaction going on between them but even a transformation from a scientific towards a technological object. These findings seem to suggest that even the analytical separation of these objects doesn't make much sense. Nonetheless, Rheinberger insists that this would be a fatal mistake, because it is the end product that counts. A scientifically produced object plays a role that is very different from that of a technological object: 'a technological object is an answer machine, a scientific object is a question machine'.[32] It is the non-technological use of technological things that affords scientific objects.

Practices Generate Agreement

Peter Galison argues for a model in which theory building, the design and construction of instruments, and experimental traditions evolve in a relation of relative autonomy to one another. At the same time, in the context of local research programmes these categories can merge, or rather, become intercalated, thus creating a new context of coherence and relatedness.[33] The 'trading zone',[34] which can be thought of as a follow-up concept to the notion of intercalation, accentuates even further the act of agreement in experimental practice. It is a place where different stakeholders interact in an experimental setting, developing their own language – or creole – and through their coordinated actions successively stabilize practices around materials and instruments.

The emphasis on social interactions in the epistemic process recalls not only Fleck's thought collective[35] and Kuhn's normal scientists working within the intellectual environment of a paradigm[36] but also brings to mind the Whewellian conception of consilience, whose success is likewise confirmed by an act of agreement making empirical knowledge a matter of convention. But where all this adds up to a mode of knowledge production that is ultimately about coordinating beliefs, shared beliefs are at best incidental side-products of Galison's shared practices. Obviously, the way agreement is constructed on the views of these authors is not the same, and nor are the kind of criteria and mechanisms of justification used. Thus one might ask whether they are all equally powerful in their assessment of the credibility of beliefs. Let us take a closer look at a fourth and final conception that contains most of the problems identified so far.

Experiments Yield Power

Rom Harré, too, is concerned about the relationship between experiment and theory, concentrating mainly on the question of how to describe more adequately the connection between the material and the represented world. Accordingly, he has pointed to the disparity between the 'persuasive power' of experiments and observations (i.e. their actual effectiveness) on the one hand and the more distant and weaker 'logical power of empirical propositions' on the other.[37] He argues that this gap needs to be filled by 'recovering the experiment' from a dynamicist point of view. This approach militates mainly against 'Latour's laboratory', which is held to be a closed system in the sense that it allows for barely any knowledge at all about the relationship between apparatus and world: everything is social and discursive, including the experimental results themselves and the technical standards applied.[38] Instead, Harré argues, we need to escape from this 'locked laboratory'[39] and find a theoretical place for the experiment based instead on the assumption that there is 'an engagement, of a sort, with Nature'.[40] When we conduct an experiment, we are engaging materially by bringing forth particular forms of representation of nature, 'namely experimental set-ups, so that by isolating fragments of nature, we domesticate it, making it available for material manipulation'.[41]

Harré proposes the notion of an 'apparatus/world complex' to recover the experiment for a material-based philosophy of science that takes into account the dynamics between constructed material objects and naturally occurring material systems, without simply settling for a distinction between the artificial and the natural. This purported distinction, says Harré, should be reframed as one 'between the domesticated and the wild. The laboratory is like a farm. It is like neither an art gallery nor a zoo – neither wholly artifactual nor wholly wild. The material setup has been tamed, rather than represented or caged'.[42]

Accordingly, the first question we might want to ask is this: what is so special about the 'apparatus' in the apparatus/world complex? Surely it is something different than simply an instrument that responds to a particular state of the material environment. Due to the fact that such responses are embedded in causal relations, we can find out something about the process occurring in nature through the changes in the instrument. Thermometers, hygrometers or photometers are good examples of experimental manipulations of this type which 'must be interpreted as analogous to states of the natural system being modeled'.[43] In contrast to instruments, the apparatus serves as a model, a simplified version of a naturally occurring material set-up. Apparatuses are domesticated material systems and, in this respect, might be regarded as similar to Rheinberger's experimental system. However, Harré develops his metaphor further by stressing that farm creatures are a domesticated version of their wild/feral counterparts. He elaborates the metaphor domesticated/wild by discussing the case of the cow and the auroc in detail, coming to the conclusion that 'domestication permits strong back inference to the wild, since the same kind of material systems and phenomena occur in the wild and in domestication. An apparatus of this sort is a piece of nature in the laboratory'.[44]

What can we conclude from this in regard to the hierarchical relationship between experiment or observation and propositions? Harré's solution is far from a harmonious synthesis purporting to offer an at once trivial and unsatisfactory resolution to a non-trivial problem, namely, the impossibility of eliminating the gap between on the one hand the world of signs and representations and on the other hand the material world of perceptible and powerful phenomena. Instead, he offers an interesting solution that comes down on neither side but envisions instead a 'both-and' – X-as-well-as-Y – structure of explanation, drawing attention to the type and the dynamics of connectedness between X and Y. In fact, his conception of connectedness draws heavily on the material (and realist) dimension where the 'powers of Nature' are accepted as given – that is, explanatory power is attributed fully to the world (and not to the social context of laboratory conditions): 'The apparatus/World complex displays the appropriate surface appearances, just in so far as the World has the power to afford them when conjoined with apparatus ... It is apparatus/World complexes that afford perceptible phenomena'.[45] Thus in a scientific laboratory the researcher's engagement with nature consists in acknowledging that the perceptible and measurable phenomena in an experimental situation are only afforded because there is an inseparable connectedness between the material set-up in the domesticated lab and in the world (wild). As long as the 'apparatus is locked into a system with Nature'[46] we are able to glean knowledge about the shared piece of nature.

The 'apparatus/world complex' is both an important and provocative contribution to the status of belief systems, including the debate on the primacy (or otherwise) of social construction. Moreover, the concept offers a fresh perspective beyond the somewhat stale artificial-natural debate. Perhaps most evidently, it also contributes to the debate over whether experimental knowledge has its own internal stability:[47] who would deny that knowledge about domesticated nature and the tamed objects themselves (the farm creatures in Harré's metaphor) is convincingly reliable?

Complementary Science and Technology?

As we have seen, the epistemological (and ontological) consequences of Hacking's famous claim 'experiment has a life of its own' are far-reaching. It touches on many fundamental conceptions in the philosophy of science, such as the separation between the contexts of discovery and justification, the question of the theory-ladenness of observation and experimentation, and the power of propositional versus non-propositional knowledge. 'The experiment also serves to create reality' may be seen as the common denominator not only in the approaches presented above but also in current philosophy of science and science studies, more generally. The relatively new philosophical field of experimentalism or instrumentalism (or scientific experimentation) has given rise to problems and questions that challenge the already troubled boundary between science and technology. This is reflected in the increasing convergence between science and technology, which can be observed also in the changing disciplinary hierarchy of the natural and engineering sciences. Historian Rachel Laudan has pointed out that 'any discussion of research that promises to diffuse the boundary between science and technology will necessarily include reflections on changing definitions of science and technology'.[48] She proposes to identify three competing definitions of technology which make it possible to map the processes of deliberation about the two contrasting modes of knowledge production, based on the question of either 'knowing how' or of 'knowing why'. The three models are 'technology as applied science', 'technology as the mirror image of science' and 'technology as the means of gaining a given end'. Laudan elaborates in detail how the three models have been realized historically and to what extent the influence of the philosophical conceptualization of technology most prominent at the time was able to (and perhaps, indeed, was meant to) play out. The model 'technology as the mirror image of science', for instance, is at work in the suggestion that the epistemologies of Thomas Kuhn, Imre Lakatos and other philosophers might be extended to describe not only scientific but also technological change.

Since the 1980s, historians of technology pursued just this programme: Edward Constant discussed the history of the jet engine by focusing on a narrative of emerging anomalies, and Rachel Laudan herself explored the consequences of a conceptualization of technology as a problem-solving enterprise.[49] At the time, this idea of the mutual referencing of science and technology also received support from the scientists' camp, which pointed to the fact that problem solving in the context of justification and methodologically weak heuristic research in the context of discovery are not mutually exclusive (particularly when modelling comes into play). Accordingly, in the 1980s social scientist Herbert Simon stated that 'we are learning a good deal, today [by working with models], about the problem-solving processes that underlie scientific creativity, and we are discovering that these processes rely basically on weak methods and heuristic research'.[50] This might be regarded as building a bridge between the science camp and the technology camp, to the extent that 'weak methods' and 'heuristic research' are usually attributed to the latter.

What can we learn from this for the description of science and technology as two different but complementary modes of cognition and practice? It seems fairly obvious that none of the three narratives proposed by Laudan really fits the current situation, which might be taken as an indication that things are actually changing as part of an ongoing process that goes by the names of 'post-normal science', 'triple helix', 'mode 2' knowledge production and any of the many versions of 'technoscience'. In her collection of definitions, it is perhaps 'technology as the mirror image of science' that comes closest to this state of affairs, in which both technology and science are not only involved in an experimental mode of knowledge production but are also expected to commit themselves to the processes of public deliberation that accompany it. The suggestion that this opening towards society also implies another type of experimentation – one oriented towards a field ideal of experimentation – will be discussed in detail later. An updating of Laudan's definition of 'technology as the mirror image of science' might yield a picture that accentuates even further the complementary, active nature of science in its relation to technology, rendering technology a bewildering 'hall of mirrors' for science – and *vice versa*. And here it seems most likely, indeed, that it is the experimental mode which is driving the increasing convergence of science and technology and which is becoming manifest in the so-called converging – bio-, nano-, cogno-, eco- (etc.) – technologies that we are encountering today.[51]

5 EXPERIMENTATION IN LAB AND FIELD

So far discussion of the 'new experimentalism' has revealed that questioning the role of the scientific experiment amounts to questioning some of the most basic tenets of twentieth-century philosophy of science. These include the distinction between a context of discovery and a context of justification, the distinction between discovering and constructing objects and, along with that, the opposition between the artificial and the natural and, finally, the separation between two dominant modes of practice and knowledge production, namely, science and technology. All this has been discussed with reference to experimentation in the lab, and here again it is the physics laboratory that has most captured the attention of philosophers of science. This is how the lab ideal of experimentation took shape and became consolidated.

For some time now, the exclusive focus on physics has come under pressure: philosophical problems concerning experimentation in the laboratories of chemists and biologists are being studied as well.[1] Strikingly, however, even this widening of perspective did not include within its scope experimental practice that takes place in the field.[2] There are only a few studies in the HPS, and some few more in STS, that have focused on the field sciences with an eye towards methodological questions and the aim of articulating a field ideal on a par with the lab ideal. One reason for this may be that field experimentation appears to be more heterogeneous even than experimental practice in the laboratory. It appears prominently in disciplines as diverse as ecology, archaeology, sociology, geology and economics.[3] However, there has also been no broadly systematic attempt by HPS scholars to compare and scrutinize these different approaches in field experimentation.

Perhaps the most striking feature of field experiments is that they deal with objects *outside*, that is, in an uncontrolled environment. Other important features of experimentation in the field include individuality, uniqueness, contingency, instability and a lack of containment or certainty. This conceptualization of the field ideal is also dichotomizing in that it distinguishes the individualizing, value-laden understanding of field objects from that of lab objects as instances of generalizable knowledge. It is for this reason, indeed, that laboratory practice can be no substitute for field experiments: 'unforeseen difficulties are found only in pilot experiments conducted under field conditions'.[4] A similar point of view is expressed in another statement from a group of biologists:

> Many major scientific and environmental sciences have arisen from a fortuitous combination of events that were associated with biological field stations ... scientists increase their chances of making serendipitous discoveries by implementing simple, well-designed projects that extend beyond conventional scales of observation.[5]

Here, a stronger thesis will be advanced, namely, that it is the field ideal of experimentation along with its historical, epistemic and rhetorical tradition that has become the dominant framework within which scientific findings and technological innovations are first tested and then applied in knowledge society. The knowledge society itself – and not the controlled spaces of a laboratory, a museum, a court or a theatre – is the stage upon which experimental design must prove itself.

Experiments performed in open spaces include social reform, medical treatment, ecological remediation and technological innovation, where each of these represents a different style of experimentation. A number of concepts that seek to label these various experimental constellations are in circulation. They include 'real-life' or 'real-world experiments', 'experimental installations' or 'innovations', 'adaptive' or 'experimental management', and prototyping. In the following, 'field experiment' is used as a generic term even while its disciplinary use in sociology or ecology, among others, remains tacitly acknowledged.

The suggestion that 'field experiment' is used in a generalized and rather more systematic sense is confirmed by initiatives such as the network for the history and sociology of fieldwork and scientific expeditions.[6] This network's point of departure is that expeditions and fieldwork have become a matter of course in many scientific disciplines, but that:

> the systematic discussion of the methodology, theory and history of fieldwork has received little attention outside the field of anthropology ... [t]he network is devoted to the study of the forms, practices and applications of scientific fieldwork as it has been conducted within a range of humanistic, social and natural scientific disciplines.[7]

Very different objects come together on the network's website: an ice core drilled by an international and interdisciplinary research team in the course of climate research in the 1970s; an imaginary sea monster, used as the symbol for a deep-sea expedition in the 1950s, that somehow escaped into non-symbolic reality by taking on a public life of its own; the 'committed collector'[8] who, during the course of his travels, collected ethnographic material on Inner Mongolia in the 1930s, mobilizing and interacting with many locals who, in turn, influenced the expedition's agenda in many ways; a fossil 'four-legged fish', that is, a paleontological object found in 1931 during a Greenland expedition that challenged the boundary between 'popular imagination' and 'scientific knowledge' from 1931 to 1955, while serving as a popular scientific magnet for political and financial support. This assortment of exemplars illustrates that scientific fieldwork covers a broad range of disciplines and scientific practices and also different degrees of engagement and interaction between the objects of research and their subjects. It also suggests that there is a certain awareness that studying the methodology of fieldwork in relation to the whole 'range of humanistic, social and natural scientific disciplines' is of utmost importance. However, scholars also agree that no satisfying synthesizing conclusions have so far been established, let alone a useful or practical systematic approach that would allow us to speak of, say, a reflexive environmental science that is pluralistic yet still operates according to a shared conceptual framework and with a collective attentiveness.

An Initial Framing of the Lab and Field Ideals

The explicit development of the notion of the field ideal in this study serves two purposes. First, it seeks to exhibit clearly the complementarity of this ideal to the lab ideal such that the intermediate forms can be described in a systematic fashion at the epistemological, methodological and institutional levels. Second, it allows us to investigate the changes in the practice of science-based experimentation in society that have occurred during the last few decades. These changes are conceptualized as a major shift from the laboratory ideal to the field ideal.

The complementary character of the two ideals can be illustrated using a conceptual scheme that sets up a series of opposites with regard to scientific practices, epistemological values and the ontology of objects. The laboratory ideal involves the design of manipulated, well-controlled, isolated experimental systems; the field ideal acknowledges their complexity, blurred boundaries and unpredictable response to interventions. Field sciences search for and find their objects outside, that is, in an uncontrolled environment. Where these objects need to be studied experimentally, field experiments go beyond mere field observation on the one hand and cannot be replaced by laboratory practice on the other hand. Essential elements of field experimental practice include the selection, reading, modification and comparison of places. In contrast to laboratories, natural places are not interchangeable sites or neutral stages on which scientific activities are played out. Since plants, animals and, of course, human beings are not inserted into these places as if they were more or less passive 'guests', and since their interactions largely define these sites, the natural places in which field experiments take place are themselves objects of knowledge. The field experiment is therefore based on a different material setting and on different metaphysical presuppositions than the lab experiment.

The following features seem to capture at least most of the differences between the two ideal types, that is, between the lab and the field as antagonistic places: 1. closed room (chamber, ship) – open space (agora, market); 2. controlled place – predominantly uncontrolled place; 3. experiment in the lab – observation/field experiment; 4. enhanced/fabricated objects – interpreted objects. The table given below offers an extended grid of these features, thereby providing an overview of the conceptions and notions discussed in the following sections. However, the most important feature is that field experiments are done in a specific variable place and, so to speak, *with* that particular place, where each of these places is the result of a singular and unique history. Historian Robert Kohler has described the field sciences as being mainly practices of place: 'Field biologists use places actively in their work as tools; they do not just work in a place, as lab biologists do, but on it. Places are as much the object of their works as the creatures that live in them'.[9] To sum up, we might say that the field ideal of experimentation is oriented towards a practice of place, where spatial openness, individuality and uniqueness, instability and contingency are the dominant features. Unsurprisingly, this stands in sharp contrast to the conceptualization of the laboratory ideal.

Table 5.1: An idea of the attributed distinctions between the lab ideal and the field ideal. As the notion of 'ideal' implies, these are ideal types in the sense of Max Weber, that is, they are 'formed by the one-sided accentuation of one or more points of view and by the synthesis of a great many diffuse, discrete, more or less present and occasionally absent concrete individual phenomena, which are arranged according to those onesidedly emphasized viewpoints into a unified analytical construct'; Max Weber in L. A. Coser, *Masters of Sociological Thought: Ideas in Historical and Social Context* (New York: Harcourt, 1977), p. 223.

Field Ideal	Lab Ideal	Source
outdoor	indoor	
uncontrolled space	controlled space	
open-space, 'market', 'agora'	enclosed-space/room, 'ship'	Gibbons et al. 2004, Barthes 1957
'practices of place'	'placeless place'	Kohler 2002
biotope	laboratope	Amann 1994
dirty real things	purified virtual things	
interpreted objects	enhanced/fabricated objects	Knorr-Cetina 1999
presence and closeness, 'passage over terrain'	absence and distance, 'steady immobile gaze'	Livingstone 2003
individuality, uniqueness	exemplarity, regularity	
instability, contingency	stability, surprise	
observation, comparison	experimenting, analogy	
contextualized knowledge	decontextualized knowledge	Lacey 2007
idiographic	nomothetic	Windelband 1884
construction of object through material intervention	construction of object through construction of observational stance	Whewell 1847, 1858

Exploring the Lab Ideal

The laboratory is the only place in society where failures and errors are explicitly welcome – and even valorized. No option that promises to enhance societal welfare is to be ignored, provided it is generated behind the closed doors of a laboratory. This is what characterizes the social contract between science and society: science is allowed, even obliged, to tinker with blueprints of reality as long as there is no risk to society. The enclosed laboratory space enables an unlimited search for facts as well as the construction of artefacts, notably for the purpose of advancing and testing scientific knowledge. Karl Popper never tired of insisting that we cannot achieve definite knowledge but only an approximation to truth or truth-likeness. The knowledge we generate draws its reliability from the trial-and-error method appropriate for assessing this tentative kind of knowledge. The underlying belief of this epistemology is that the power of critique is strong enough to uncover errors and failures in our theories.

This turns the lab into a place where anomalies are cultivated and where new facts or objects are cheerfully sought and accrued without being called into question by society. Accordingly, one important condition of the lab is its character as a container, as a secure, enclosed space in which instruments and machines are applied to afford scientific phenomena and objects that can then become inscribed in a theoretical language. The following sections provide some snapshots of the process by which the experiment came to be liberated from the literal walls of the laboratory. The first phase of the process of liberation can be seen in the expansion of the experiment as a cognitive tool using the entire laboratory as a cognitive space – and not just the experimental devices on a lab bench. In order to track this liberation, we need first to reflect on general criteria of experimentation as they have been derived (by Martin Carrier) from physics and then see what happens to them as they are adapted (by Kristian Köchy) to biology.

Philosophical Criteria for Scientific Experimentation

As was outlined in the preceding chapter, the conceptualization of experimentation in the laboratory has changed considerably over the last few decades. In the following, philosophically sound criteria for scientific experimentation will be presented. This facilitates a close look at the intertwining of programmatic (or even ideological) and theoretical aspects of experimentation.

According to Martin Carrier, experimentation in the lab can be characterized by the three criteria of isolation, intervention and completeness.[10] Of these, *isolation* is the most familiar, as it concerns the requirement to separate the experimental system from the surrounding environment in such a way that it becomes possible to observe individual parameters and their effects on the system. Once stabilized and established, the setting of instruments and methodological operations also guarantees the reproducibility of the experiment.

Carrier's second criterion is *intervention*, which, in conjunction with isolation, has the aim of producing causal knowledge. An intervention happens when an experimentalist induces a particular event and the experimental system reacts in one way or another.

The third criterion, *completeness*, presents the most challenging requirement. Ideally, the experimental system is to be designed in such a way that it enables a systematic variation of interventions by any parameter that is regarded as being relevant to the investigation of the system. It is again the enclosed and controlled space of the laboratory that allows for a comprehensive analysis of these parameters. To be sure, this idealized situation contrasts sharply to the conditions outside or in the real world, whether in nature or in society. Here, only a certain constellation of parameters is typically realized in a particular range of dimensions, such as a particular variation of moisture or a limited number of possible partners for mating. Only the artificially enhanced environment of the laboratory puts an optimized range of such parameters at the disposal of experimental manipulation.

Challenging the Criteria: Experimentation in Biology

Kristian Köchy[11] explores the historical genesis of the power granted to the experiment. He offers a scheme that generally adopts Carrier's three criteria but gives them a critical normative twist and thus lays the groundwork for the argument that experimentation in biology requires a different framing. Köchy's first notion is not isolation but *separation*. He points to the necessary assumption of any laboratory experiment that it is possible, in terms of significant propositions and theories, to examine a natural situation in isolation. Köchy points out that this criterion serves a reductionist ideal and presupposes the application of an analytical procedure.

Similarly, Köchy's second notion of *control* takes up the theme of intervention and emphasizes that it manifests scientific rationality in that by means of experimentation a certain kind of effect can be produced that indicates a causal relation. For theoretical purposes and also in practice, the experimental design is only satisfying if this effect is produced reliably so that the experiment's reproducibility can be attributed to the causal relation. Only through this attribution can the requirement of intersubjectivity or the expectation of applicability be satisfied. The underlying assumption that ties causality to reproducibility is that there is homogeneity and lawful regularity in nature. It is then a matter of philosophical interpretation whether one speaks of the discovery of lawfulness in the laboratory or its antecedent construction. Either way, the construction of reproducibility goes along with the stabilization of a scientific object under various and changing conditions – in other words, a phenomenon in the laboratory becomes transformed into a scientific object.

Finally, Köchy reinterprets the notion of *completeness*. The scope of relevant parameter variation is determined by the distance between experimenter and object. Köchy makes the point that in the classical conception of science, techniques of creating distance and separation are used to make things look objective; they reflect a particular 'form of authority and priority'.[12] Köchy is not satisfied to formulate these notions as general criteria of experimentation along with those of Carrier. He goes on to argue that the conception of this ideal of experimentation needs to be refurbished if it is to include the specific conditions of experimentation in biology, let alone the diverse experimental modes in the various disciplinary and research fields of biology. His proposals for 'cutbacks, modifications, or re-phrasings'[13] of the idealized criteria for experimentation will be discussed and commented on in the following.

The requirement of separation or isolation runs up against the problem of connectedness in biological systems: how we can even know which relationships or parts of a system are relevant to the system and which are not? Philosophers of science clearly acknowledge that sorting out which variables are the relevant ones and how to avoid extraneous influences is a special difficulty encountered in biological experimentation.[14] It appears that experimental design in certain fields, such as molecular biology, microbiology or traditional physiology, is rather successful. This depends, of course, on what counts as success: as long as it is a matter of identifying a metabolism cycle, an essential metabolic step or even a new class of metabolites, there is a good chance that the problem can be conceptualized well enough and investigated empirically in an appropriate experimental setting. But even in such seemingly simple cases, methodological complexity and the difficulty of comparing between different theories make themselves felt.[15] This was shown in the context of the problem of oxidative phosphorylation, the so-called ox-phos controversy.[16]

Another example is the problem of the metabolism of cancer cells, which eventually became transformed into a problem of identifying and representing ribonucleinacid. When the researchers finally described their model of RNA as a 'mechanism of incorporation of labeled amino acids into protein' they effectively paved the way for a research programme of 'genetic engineering'.[17] There are myriad examples of this kind of successful problem solving and knowledge production in the many fields of biological research. However, as soon as the laboratory is left behind things get more difficult: separation without isolation – or with weak isolation only, as in the case of experimentation with the release of genetically modified organisms – heightens the difficulties of separating out the effects of the intended intervention from noise in the environment. Research fields confronting these difficulties are often classified as weak at best, as can be

observed in the philosophical discussion on laws/law-likeness and the controversy over the status of specifically biological regularities. This will be considered
with specific reference to field experimentation in the next chapter 'Exploring
the Field Ideal' and in the discussion of plurality in ecology in Chapter 10.

The problem of manipulation brings with it the question of what counts as
an experiment and what counts as an observation. In contrast to astronomical
or physical phenomena (such as the structure and dynamics of sunspots or of
the tides), most biological phenomena are not scientifically observable without some kind of minimal intervention at least: plants and animals cannot be
observed without entering the field, whereas observations made with a magnifying glass or a microscope require that the living object assumes a somewhat fixed
position. When a wild animal is tagged to monitor its movements or those of
the population, or when we want to know about the decomposition processes of
oiled sediments, we need to intervene and might even be forced to change the
natural processes to an extent 'that has to be diagnosed as the loss of "life" – as
"dead". There is nothing comparable to this situation in the natural sciences'.[18]

Two things should be noted at this point. First, in biology observation and
experimentation also involve intervention and thus some degree of manipulation
of the object under consideration. This intervention goes beyond the construction and manipulation of the instrument – of a microscope, for instance: 'you
learn to see through a microscope by doing, not just by looking'.[19] Second, the
organization of living organisms does not respond to interventions in a continuous way but rather in a discontinuous way and usually irreversibly. Organisms
(or cell cultures) are only resilient to a certain extent within a certain range of
values and might suddenly change their organization in a destructive way if a
critical value or a time limit is reached. This phenomenon is rarely encountered
in physical systems and therefore requires special attention when designing
and reflecting on biological experimentation – processes often attended also by
moral, practical or legal considerations.

The reproducibility of an experiment might be seen as being an even more
difficult criterion to fulfil in biology. This is mainly due to the importance of singularity that plays out regularly in the domain of biological objects: no cell, no
organism and no ecosystem is exactly like another, so the first step in a biological
experiment might be to reduce this heterogeneity, for instance, by breeding cloned
mice or cultivating particular cell lines. Complete homogeneity of the biological
material is therefore a criterion that can only rarely be achieved, if at all.[20]

However, none of this is surprising given that one of the hallmarks of biological theorizing is to insist precisely on this: evolution is based on and advances
singularity and individuality; it generates the diversity and variability of organisms, of the stages of their life-cycles, and also of assemblages of organisms.
Modifications and errors – nature's experiments – occur at all levels – in the

genes (and other molecular structures) and in the population; they also act, of course, on the individual. Natural selection describes procedures and processes that depend on initial conditions; it is regarded as a natural law, yet it is accorded an exceptional status: it is not universal, applying only to specific systems, and it allows only for limited predictions.[21] All these factors serve to explain why biological research is case study research, which is not a deficient style of experimentation but a necessary and ultimately thoroughly suitable strategy to learn about biological objects.[22]

Like reproduction and homogeneity, control and distance are two sides of the same coin. Controlling an experimental situation implies not only stepping back and adopting an attitude of a distant ('objective') observer (a 'God's eye view') but also accepting that reduction and simplification are necessary to receiving any response amenable to evaluation. Control implies the creation of an artificial environment and the exclusion of parameters that have been deemed to have no relevance or only secondary relevance. This leads back to the problems already discussed under the heading isolation/separation and that can be identified as the specific difficulties of biological research, that is, the difficulty of dealing largely with autonomous and self-organized research objects. Controlling and distancing the research object are closely related. In a paper on the moral economy of science, Lorraine Daston discusses how emotions and values are subject to purification in order to realize the ideal of scientific objectivity. She concludes that '(t)here is excellent evidence that the moral economies of science derive both their forms and their emotional force from the culture in which they are embedded – gentlemanly honor, Protestant introspection, bourgeois punctiliousness.'[23] The difficulty of attaining this distance and control in biology is underscored by Theodore Porter's argument that the interpenetration of authority and objectivity has been mediated by the power of numbers and the increasing significance of quantification since the nineteenth century.[24]

Distancing the subject of inquiry is not only a popular and historically deep-rooted attitude in biology and the natural sciences in general. It is also regarded as an ethical issue, or rather ethos, in the sense of suppressing affection or concern. It might be seen as part of the changing relationship between science and society that this ethos became a subject of various political and philosophical controversies. The focus of one such was the apparent disappearance of the scientist's body.[25] Another strand of debate developed around scientific agency and the call to render it visible rather than hiding it. Finally, the issue of scientific responsibility for the 'unnatural' constructs, or 'monsters', created in the laboratory is also related to the question of distance and the power of objectivity.

Experimentation in the Lab: Establishing 'Enhanced Environments'

A different approach to science and experimentation was developed by so-called laboratory studies.[26] These provided a new perspective on the production of scientific knowledge in 'these strange places called "laboratories"',[27] one that proved to be so innovative and productive that it came to be known as the 'practical turn',[28] or the 'practice turn in theory'.[29] Laboratory studies focused mainly on scientific practice as a cultural activity of a particular tribe, the community of scientists, and probed the perspective of 'the traveller in exotic lands', where laboratories are places that are observed using the 'objective yet compassionate eye of the visitor from a quite other cultural milieu'.[30] This other cultural milieu was most often the one of sociology of science (or sociology of knowledge), social anthropology or, in some cases, art history. One notable characteristic of this approach is that it is susceptible to the particularities of laboratories as specific places, be it as epistemic places, social places or aesthetic places. This importance of place and of the specifics of the sites of scientific research was also emphasized by Galison and Thompson, who referred to the laboratory as the scientific dwelling: 'Place and dwelling held a preexisting attachment to the world, an organic connectedness of the home to the landscape'.[31] In their book *The Architecture of Science* they demonstrate nicely the diversity of architectural sites for scientific research and how this situatedness influences the production and validation of scientific knowledge, including experimental knowledge (see also the next chapter).

Another important characteristic of these laboratory studies is their emphasis on the constructedness of scientific knowledge along with the making and constructedness of laboratory things – a position that entails opposition to some philosophers' ontological realism. Here, the artificial or hybrid character of processes and phenomena in the laboratory refers back to Francis Bacon's experimental philosophy. When Latour talks of the laboratory as a 'solid lever'[32] or 'leverage point',[33] this recalls Lichtenberg's comment on Bacon's *Organon* as a heuristic hoist (see Chapter 1), emphasizing the heuristic art of finding and the proximity between the discovery and the invention of things.

The following section examines the role played by the laboratory and experimentation in transforming natural things into enhanced scientific objects. In contrast to ordinary, naturally occurring, unprocessed events or things, these objects are 'enhanced' insofar as they are able to prove or even embody scientific facts. Here, the approach of Karin Knorr-Cetina appears particularly promising, especially her conception of an enculturation of natural objects in the laboratory: 'The power of the laboratory (but of course also its restrictions) resides precisely in its enculturation of natural objects ... The laboratory subjects natural conditions to a social overhaul and derives epistemic effects from the new situation'.[34]

The Laboratory as a Cognitive Space

Proponents of laboratory studies hold that it is not sufficient to elucidate science by focusing solely on theory and experimentation, because the scientific institutions and social networks in which the experiments are designed and conducted are not taken adequately into account. Ignoring the significance of the social interactions entailed in theory and experiment likewise obstructs the view on daily life in the laboratory. This was identified as a problem because the 'luxurious chaos of mundane relationships', as Knorr-Cetina put it, intervenes in the construction or manufacture of the scientific object and is thus also involved in knowledge production.[35]

For this reason the scope of consideration was widened from the methodologically limited experiment to the laboratory as a cognitive space. This enlargement of the research object from the experiment to the lab gives equal prominence to the technological fabrication of scientific objects, their symbolic and political configuration, and the strategies deployed to distinguish one's own from competing laboratories.[36] It is this interdependency of different practices in the lab that is brought to light by laboratory studies.

Karin Knorr-Cetina conceptualizes the laboratory as a place where things are gradually shifted from a natural order towards a social order, as belonging to the actors and their environment. It is a place where scientists capitalize on their socio-cultural conditions and positions, doing so by using the interpretive flexibility and plasticity of natural objects. This interpretive flexibility had previously been emphasized by Ludwik Fleck in his discussion of the direct perception of form, the 'Gestalt-perception',[37] and of seeing-this-way-or-that-way, describing the cognitive process of the trained gaze that comprehends matters at a glance in, say, microscopy.

> Direct perception of form [*Gestaltsehen*] requires being experienced in the relevant field of thought. The ability directly to perceive meaning, form, and self-contained unity is acquired only after much experience, perhaps with preliminary training. At the same time, of course, we lose the ability to see something that contradicts the form. But it is just this readiness for directed perception that is the main constituent of thought style.[38]

Thus laboratories benefit from the circumstance that objects do not represent resilient or 'recalcitrant' entities, but that they prove to be plastic and flexible.[39] Laboratory scientists do not manipulate and interact with objects as they might occur in nature. Instead, they are handling optical, acoustical or electrical traces, confronting components and fragments, engaged in purification and breeding. Lab scientists need not deal with natural objects as they are and they must not work on them where they occur in their natural environment. Instead, research objects are brought into the laboratory in order to be investigated on its terms and under its conditions.

If, for instance, we decide to use a mouse as a model organism, we cannot just descend to the basement or walk in the forest and try to catch an unlucky individual house mouse or wood mouse. The model system 'mouse' that enters through the door of a laboratory most probably originates from another laboratory specializing in mouse breeding and offers a whole shelf of different models, each cultured for a particular purpose, all of them products of a fusion of many different cultivated breeds and wild forms, natives from Europe, Japan or the US.[40] When it arrives in the laboratory, the 'wished-for mouse' is further cultivated to be available as a model system at all times and in sufficient quantity. For this purpose a 'laboratope'[41] – as opposed to the biotope – is constructed: an entirely artificial environment in which every input and output is controlled – light and heat, mating and feeding, life and death.[42] Indeed, the mouse itself might become an environment, as when its reproductive organs are used to produce transgenic mice. All these transformations can be described as a process of enculturation[43] in which the corporeality (*Leiblichkeit*) of the mouse vanishes. This material transformation is the first of a series of displacements from the model system 'mouse' into various media of representation, such as gel electrophoresis plots, gene maps, diagrams or a PowerPoint presentation.

Accordingly, the potency of the laboratory consists in its power to manipulate natural processes and to simultaneously exclude the kind of uncontrolled nature that lurks outside the laboratory's walls. This is why the laboratory has been called an 'artificial space through and through',[44] an 'enhanced' environment.[45] This reframes and underscores what Gaston Bachelard had previously noted: 'we will have to demonstrate that anything that is made by humans relying on a scientific technology does not exist in nature and is not even a natural consequence of natural phenomena.'[46]

Lab-field Co-production

The question of the boundary between the lab and field sciences is a conspicuous one in the cultural geography of contemporary science, particularly of the biosciences. It refers to the issue of how the knowledge produced is connected to the real world, to physical places and to the socio-political context. It thus reflects on the notion of scientific practice and its scope: is it a predominantly observational or experimental practice? Does it follow the lab or field ideal of experimentation? And what type of theoretical knowledge does it advance, one that is rather case-oriented and produces local knowledge or a different kind aimed at establishing universal laws or law-likeness? All this ultimately leads to the question of the credibility, prestige and standing of a discipline or field of research.[47]

There appears to be agreement that the lab-field boundary in biology is of rather recent origin. There is more disagreement as to what 'recent' means and how to date the origin: Kohler holds that it is probably no older than the mid-nineteenth century, when laboratories outgrew museums and herbaria as the premier places of modern biology.[48] Indeed, the concept of the 'field' can be regarded as a by-product of the so-called laboratory revolution of the second half of the nineteenth century.[49] Kohler points out that 'the categories of field and laboratory were co-invented and are mutually (and changeably) defining: like matter and antimatter, town and country, good and fallen angels'.[50] Once the scientific geography was dominated by laboratories,[51] the long-standing experiential techniques of observation and comparison lost prestige and were downgraded to 'second-best practices'.[52] This is a consequence of the creation of the urban 'factory-laboratory', as Gaston Bachelard called it, with its attendant new knowledge-producing tools and technologies. Accordingly, he characterized chemistry as a science with a technological dimension that seeks the 'active transformation of matter'.[53]

This principle of a growing technological dimension in the process of producing knowledge was also observable in the process of increasingly understanding life phenomena during this extremely innovative 'laboratory revolution': 'physiology gained stature as one of the century's *Leitwissenschaften*' and 'increasingly observed, dissected, measured, stimulated, registered, and graphically recorded in laboratories'.[54] The life sciences were caught in the middle of this 'revolutionary' process that restructured the landscape of scientific practices and knowledge and finally also engendered the lab-field boundary within biology.

A short story about the social and scientific decline of the *Musée d'Histoire Naturelle* in Paris due to this 'laboratory revolution' might serve as an illustration here. In the first twenty to thirty years of the nineteenth century the museum was the leading scientific institution worldwide in terms of the amount of material collected and the systematic-theoretical achievements. The arrangement of the collections and biological theory building were directly connected to one another, while the museum formed the intellectual centre of indoor and outdoor scientific activities alike. The inseparability of theory and collected material became visible in quite literal terms when Lamarck and Cuvier clashed over what was the correct zoological system, displaying their models in two different halls of the museum, Lamarck in the *Galerie de Zoologie* and Cuvier in his *Cabinet de Cuvier*.[55] From the 1840s onwards, however, the museum progressively declined in influence and prominence for different reasons, the most serious of them being the scientific rivalry with the emerging experimental sciences. The laboratories of the latter were situated at the university, and the state authorities decided that this was to be the new educational structure that a modern state needs. The universities and, with them, the scientific laboratory became the showcase of the Third Republic.[56] The budget of the museum was reduced, positions reallocated and collections neglected, the ruinous state of the buildings reflecting the situation of the only now self-identified 'field sciences'.

As for the recent origin of the distinction between lab and field in biology, Dorinda Outram argues that the boundary between indoor and outdoor science was already the subject of debate among biologists in the late eighteenth century or even earlier.[57] This is not necessarily in contradiction to Kohler's claim regarding the genesis of the lab-field boundary around the 1850s because, as Kohler himself argues, the distinction between laboratory and field does not coincide with the ones between armchair and expeditionary science, between indoor and outdoor science. Unlike the contrast between lab and field science, the indoor and outdoor sciences pursued rather complementary interests in that they both referred to the same fundamental conception of nature. There was also agreement about the importance, and even primacy, of fieldwork in that it provided the material for the indoor biologist – meaning not just plant and animal specimens but also direct, 'authentic' experiences, in written and oral form. By contrast, indoor biologists had the available instruments and the knowledge enabling them to deal with the ever growing bulk of material by bringing it together and forging connections. The indoor biologists had the analytical tools to create order from an overwhelming and chaotic world. As opposed to the co-produced antithetical laboratory and field sciences, '[c]loset science and fieldwork were two ways of doing natural history, not two distinctly different kinds of science'.[58]

David Livingstone provides a somewhat different analysis of the 'open-space naturalists' and the 'closed-space naturalists', as he called the fieldworkers and the laboratory naturalists, by emphasizing their different cognitive styles.[59] He develops his conception mainly through reference to the controversy between Georges Cuvier and Alexander von Humboldt, the first a famous museum naturalist, the latter a famous (if not *the* most famous) travelling naturalist of his time. Cuvier marks the difference by pointing out that fieldworkers can only make 'broken and fleeting' observations[60] in contrast to the worker at the bench, who has the time and the perfect place to focus on samples, to analyse them, and to develop in the end a reliable conception, integrating in one system the new and old results, the items freshly discovered and those already known. For the fieldworkers it was presence and closeness that validated their claims, whereas for the stationary naturalists it was absence and distance from wild, unenculturated nature. The former generate power from their '*passage over* terrain', the latter from their 'steady and immobile *gaze*'.[61] However, as Livingstone points out, there is still a balance of power between the two types of naturalists, the one providing the material, the dirty but real thing, the other one the purified but virtual world of a natural system. The setting changed, however, when the laboratory world gained more and more autonomy and was able to reproduce itself to a significant degree. The real things from open space now came to be substituted by virtual things in the laboratory's closed space which perfectly fulfilled the requirements of this virtual world. However, these virtual things (transgenic mice, for example) are far less viable (if at all) in the real world. The real things in open space are now constructed as denizens of a different world. Accordingly they are to be described by a different ideal of scientific practice and theory – the field ideal.

6 EXPLORING THE FIELD IDEAL

And the wind bustled high overhead in the forest top. This gay and grand architecture, from the vault to the moss and lichen on which I lay, – who shall explain to me the laws of its proportions and adornments?[1]
>   Note in a journal written by Ralph Waldo Emerson, 11 April 1834

We have seen that working in the lab and doing fieldwork are two activities closely related to one another throughout the history of science and that conceptions of lab and field are co-produced. This can be highlighted by juxtaposing them as ideal types, the lab ideal dealing with controlled as opposed to uncontrolled places, where methods of intervention rather than those of observation are used and where knowledge production happens in closed rather than open spaces. As the ideal types are complementary to one another, a closer look at the experimental practices in the laboratory will help to also reveal the characteristics of the field as a site for scientific experimentation.

Laboratories are places designed for experimentation. They were created specifically for a group of researchers to inhabit – a dwelling place in which research programmes that imply specific experimental systems are pursued. Labs are built and controlled places and, in spite of their highly functional design, tinkering is a dominant practice that goes on there. Labs are simplified and indeed standardized places in which decontextualized knowledge[2] is produced and environmental variations excluded. The main purpose of simplicity and sameness is to enable the criterion of reproduction to work in such a way that we trust experimentation more than other ways of producing knowledge. The notion of the lab as a placeless place captures this quite nicely:

> We take placelessness as a diagnostic of universality ... Being everywhere the same (more or less), labs give us no visible reason to suspect that knowledge produced in any lab is not universally true but merely true for some people in particular.[3]

Field practices, by contrast, are clearly practices of place, regardless of the disciplinary perspective of the researcher, who might be an ecologist or a geologist, an anthropologist or a palaeontologist. Selecting, monitoring and comparing places, reading and inscribing traces, collecting, tagging, and perhaps preserving specimens are the essential elements of field practice. Observation is the dominant cognitive mode of the field. It is structured in an experimental fashion, however, and as such is far from being a passive or neutral activity. This tends to blur the boundary between mere observation and field experiment:

> Like experiment, observation is a highly contrived and disciplined form of experience that requires training of the body and mind, material props, techniques of description and visualization, networks of communication and transmission, canons of evidence, and specialized forms of reasoning.[4]

As has been suggested above, the field experiment needs to be clearly distinguished from the experiment in the lab because it can be identified as a rather weak intervention in the system. As such, it can be regarded as approximating more to strategic observation. This might be comparative to changes that are *selected* rather than *introduced* by the experimenter. But as was also shown above, the tendency to identify experimentation with strong interventions in the designated space of the laboratory owes to the desire to restrict it to a safe place where the risk of failure can be contained. This desire exists in a relationship of tension to the Baconian spirit of experimentation that extends to society and the advancement of learning generally. The negotiation of this tension is ultimately a matter of ideological boundary work on the apparent contrast between passive observation and active experimentation. This boundary work appears to have been settled in favour of the view that strategic observation and social learning require an experimental approach in the fields of technology and economy and indeed in society more generally.

Accordingly, a key element of the field ideal is that it involves an experimental type of observation. Such strategic observation constitutes more than the simple acknowledgement that any attentive perception involves an active observer. Instead, it concerns the specific way in which observation can involve interventions even where the observer has not physically induced that intervention (as in Whewell's isotidal maps, for example) or where the intervention lacks the controlled specificity of a parameter variation within a laboratory. A maxim such as 'observation excludes intervention' in contrast to 'experimentation always implies intervention' obviously does not apply: although the latter is certainly correct, the former does not hold for scientific observation. Accordingly, the difference between observation undertaken in laboratory and in field experiments is a difference in degree of intervention but not a difference in kind – much like the difference between the life sciences and the physical sciences.[5] This difference in degree of intervention corresponds to a difference in degree of how active the 'active observer' is.

Another key element of the field ideal as opposed to the laboratory ideal is that natural places are themselves subject to scientific scrutiny and research practices. Plants and animals in the wild are not passive 'guests' in a particular place. And they are not like model organisms that form part of the experimental system in the lab rather than inhabiting the lab as a place. Organisms in the field alter the places they inhabit and are altered by them – this requires at the very least the biological concept of a modern organism in its environment. It follows that field biologists have to include place in their investigations, because it is part of their object of study. It is 'Nature's experiments' that can be recorded and deciphered in these places.

Unlike labs, places in the outdoors are also uncontrolled in that they are subject to multiple uses: hunters, fishermen, artists, tourists, environmentalists, urban dwellers and farmers all have their own ideas of what a natural place is and what it is for – ideas which may often clash with those of ecologists or other scientists. It is a commonly held view that this social diversity compromises the scientist's (for example, the field biologist's) credibility and social standing:

> In a world of specialists we are obliged to judge the value of knowledge by the social qualities of the knowers, and we judge their qualities in part by the social space they occupy. Knowing that access to laboratories is restricted to a certain kind of person is a powerful symbolic guarantee of credibility (look for the white coats).[6]

Given this degree of prestige and credibility attaching to laboratory science, it is not surprising that biologists (and, of course, other field scientists too) have tried to transfer laboratory methods into the field and to import the field into the lab. It is at this point that the features of the two ideal types merge and where epistemologies and social and institutional cultures become patchy. These dappled places might also be called 'dirty places': they exist somewhere in between the ideal types of lab and field, neither fulfilling the criteria of ideal lab experimentation, as discussed above, nor corresponding to a pure field ideal. Such places include museums of natural history and anthropology, zoological and botanical gardens, aquariums and mesocosms inside and outside the scientific laboratory, more or less controlled greenhouses of a more or less impressive size, such as Biosphere 2 in Oracle, Arizona (US) and the Eden Project in Cornwall (UK), but also so-called living labs – 'open-innovation ecosystems' – usually situated in an urban environment and set up to test the prototypes of technoscientific products and lifestyles. All these places are, in many respects, hybrids, so it is no big surprise that the large majority of places where research takes place outside the laboratory are such dirty places.

Placing Place

Notions of place, space and the local have proliferated in recent processes of deliberation on the recalibration of science, society and technology – indeed they have almost come to be overused or hackneyed concepts. Michel Foucault's clear-sighted analysis that twentieth-century thought is characterized by its pre-occupation with space seems to have been proven correct.[7] Topographical and spatial notions recur consistently, both as theoretical tools and as descriptors. So it is all the more astonishing that there are relatively few studies on the concept of place itself, on what distinguishes place from space and what marks one place off from another. What seems to be lacking is a careful analysis of the many different meanings of place and the spatial, 'a topography of place as such'[8] that would lead to a 'recognition of particular characteristics of space and ... a politics that can respond to them'; what is needed are 'reorientations concerning both place and space'.[9] The emphasis in this section on placing place seeks to address this conceptual neglect and to highlight the blind spots it produces. These blind spots hamper our understanding of place especially in relation to the semantics of difference and structure – which is particularly relevant, given that place is thought to play a significant role in theory. There may be a number of reasons for this neglect (or avoidance). Without going into too much detail here, just two will be singled out: first, the seeming clarity of the everyday use of 'place' and, second, a suspicion of universalizing tendencies. The latter, in particular, clearly works totally contrary to the purposes envisaged in exploring the topographies of relations between science and society and in examining the contract between science and society with particular regard to the speculative moment of the (de) liberation of the experiment. The inclination to consider places only in their particularity stands in the way of *siting* the experiment within the topography of the various ways in which science and society are related.

Especially since the 1980s a great deal of interest has developed in singularity and particularity of context. This has found expression in phrases that evoke notions of place, space or locality.[10] Not least among these is the variety of '-scapes' that have appeared in scientific as well as in more popular literature. Examples include labscape, geoscape, deepscape, soundscape and, most curious of all, spacescape – all of them referring back to the 'scape' we first knew from the word 'landscape'. Dating back to the very beginning of the seventeenth century, landscape, or *Landschaft*, is etymologically derived from 'land' and 'ship'. And, if we follow Roland Barthes's interpretation in 'Mythologies', 'ship' in turn is 'the ultimate emblem' of closure.[11] 'To like ships is first and foremost to like a house ... [A] ship is a habitat before being a means of transport. And sure enough, all the ships in Jules Verne are perfect cubbyholes'.[12]

In contrast to closed '-scapes', places are open – at least, this is what we learn from the Latin word 'platea', meaning a 'broad way' or 'open space'. Place, *Platz* and *piazza* are all closely related to this meaning, and suggest a bounded openness. They contain the notion of an open but well-defined area, most often the square or marketplace within a town. This notion of place differs markedly from the concept of space as a homogeneous, extended realm that is amenable to a purely mathematical understanding and within which such 'places' may themselves be situated. This leads to a tendency to take space paradigmatically to be a concept of physics, to treat place as secondary to it and, at worst, to view any other concepts of place or space as purely psychological. Thus it has become common practice to subsume the notion of place within the concept of space.

However, the interesting question seems to be this: how can we take account of the idea of place as a bounded but open realm with a universalizable character of its own, as for example, the *agora*,[13] or marketplace,[14] in the debate over mode 2 knowledge production? In ancient Greek democracy, the *agora* is a marketplace of ideas located just outside the gates of the parliament. It is a place characterized by the absence of determinate power, offering a relatively unbounded space of discursive possibility that complements the spaces of technical and historical possibility. Of places like this, one might say that they designate a type of experience or a type of human relatedness to other humans or their environment.[15] This distinguishes them from merely local and parochial places on the one hand and from abstract or physicalist space on the other. Such conceptions of place have been developed in disciplines as varied as environmental philosophy, geography, political ecology and ecology, sociology and social anthropology.[16]

Yet it seems that there is more in this alliance between place, experience and thought that might provide an important clue for understanding the distinctiveness and structure of place. What if we look at this alliance the other way round? In other words, rather than conceptualizing the relationship between human beings and their world as one in which the former impose meaning onto an otherwise objective, physical structure, we might see that 'understanding the possibility of human being – or meaning – is just a matter of understanding how place as such is possible. Understanding human being and understanding place are then one and the same'.[17] In his paper on 'finding place', James Malpas holds that places themselves are structured and that the way in which structure constrains the study and delineation of places provides an important model for our understanding of human beings as well as of human being.[18] It enables us to ask how subjectivity is structured when it is interrelated with other things and beings. What follows from this is a reframing of the process of understanding things – and therefore a reframing of the process of producing knowledge about these things: it is more important to investigate their interconnections, or worldliness, than to isolate their dependencies or pursue their reduction.

There is nothing really new in this – we encounter this sentiment throughout the history and philosophy of science of at least the last 200 years, in the writings not only of Johann Gottfried Herder and Martin Heidegger but also of Ludwig Wittgenstein and many others. However, what we might identify as being characteristic of recent debates is that the deliberation of place and space arises from a sense of disconnectedness, and that we can re-evoke connections and connectedness by turning our attention to things and places[19] – both of them attractive and promising candidates for exploring a dappled, hybrid, dirty world that is full of value.

As mentioned above, general ideas about particular places have been picked up and elaborated in philosophical and scientific reflections on the environment, especially in biology and ecology. Some examples may serve to show how these also delineate a specific understanding of the human being. In tandem with his conception of an active relationship between an organism and its environment, Jacob von Uexküll introduced the concepts of *Merkraum* (perceiving space) and *Wirkraum* (affecting space), emphasizing the perceptual and active characteristics of a particular, localized configuration of organism and environment.[20] With his 'deep ecology' approach, philosopher Arne Naess urged greater attentiveness to humans' connectedness to places, particularly to their homeland or home region (*Heimat*), because places teach human beings about values and connectedness, and this in turn teaches them respect for and attentiveness to the places of other creatures, be they human or non-human beings.[21] A phenomenological approach is adopted by Gernot Böhme and Gregor Schiemann,[22] who propose that places should be observed through their history and conceptualized as a texture of relations, while Christoph Rehmann-Sutter[23] and others have argued that, rather than merely addressing conceptions of place, a discourse on environmental ethics might itself best be *structured* by such conceptions.

Viewed in the light of lab-field relations, each of these approaches may provide helpful clues when exploring the kinds of place discussed in the remainder of this section, namely, (1) place as a counter-concept to the Cartesian concept of space, (2) place as an ambiguous concept expressing a simultaneously open and bounded space/location, (3) place as a structure where things and beings are interconnected and 'become grounded' and, finally, (4) place as an arena for agency and action to create connectedness.

Experimentation in the Field

Again another element is required to make the field experiment work at all, namely, that a group of scientists must seek to engage in a process of deliberation on a problem or object in the field. This must not necessarily happen within a specific scientific community or discipline; rather, scientists might be attracted by a problem or object for a while, work on it together, and then return to their respective institutional and disciplinary settings. While this sounds obvious enough, the notion of group deliberation needs to be explored, both in regard to the agreement that is presupposed by or achieved in deliberation, and in regard to the social interactions that are required in field experiments.

First, scientists do not necessarily need to agree on, say, basic epistemological issues or even background assumptions (regarding different conceptions of nature, natural philosophies, and so on). They simply need to get together to work on their local problem. There are different views about whether this working together should be called cooperation or collaboration (the latter places greater emphasis on consensus).[24] Susan Leigh Star and James Griesemer explicitly reject the implicit naturalism of the term cooperation, which they say suggests that the common goals of action are somehow grounded in the nature of the scientific enterprise. They thus consider cooperation as a 'common myth' about science. Instead, because

> new objects and methods mean different things in different worlds, actors are faced with the task of reconciling these meanings if they wish to cooperate. This reconciliation requires substantial labor on everyone's part. Scientists and other actors contributing to science translate, negotiate, debate, triangulate and simplify in order to work together.

In so doing, scientists produce representations or objects, and these 'representations, or inscriptions, contain at every stage the traces of multiple viewpoints, translations and incomplete battles'.[25]

It might appear that it is not enough to require as the presupposition for field experimentation that researchers engage in a process of deliberation on a problem or object, since all scientific activity (be it theory building or the development of methods) ought to be aimed towards achieving intersubjective agreement – otherwise it constitutes merely a para- or pseudo-scientific theory, an artwork or simply a private belief. The main figure identified with this appeal for the possibility of intersubjective agreement is Karl Popper, who described it in his book *The Open Society and its Enemies* (1945) as resting on the existence of open criticism on the one hand and 'on the public character of the experiences (observations, tests) to which scientific theories must be answerable' on the other.[26]

However, this appeal does little more than underline the fact that social interaction is necessary to make an experiment work. When it comes to the question of how much agreement is required to embark on an experiment together, though, this might actually differ between laboratory and field experiments: while lab experiments have an unspoken consensus about experimental design literally built into their architecture, field experimentation may well require a greater degree of social connectedness in order to compensate for the absence of the laboratory walls. But taking on board Star and Griesemer's point,[27] this social connectedness need not take the form of agreement.

This greater social connectedness appears to be counterbalanced by the fact that working outdoors is often a one-person activity. This would suggest that scientific observation in the field is primarily an activity undertaken by a solitary middle-class subject who enjoys 'nature' or landscape. Indeed, this image of the field experiment has been conjured in debates on the epistemology and the history of geography or landscape ecology. However, it is a misleading image that underestimates the epistemological specificity and social complexity of field experiments. It underestimates epistemological specificity because, in the solitary observation of 'nature', a nocturnal encounter with a wildcat, for example, adds just one more piece of sense data to a precisely planned grid that serves to fulfil specific statistical settings. It underestimates social complexity in that this solitary nocturnal encounter is the result of a continuing process of deliberation between different stakeholders – other researchers in the team, forest officials, perhaps eco-activists, the project sponsors and so forth.

Landscape and Idiographic Methodology

A fourth key element of field experimentation concerns its reliance on an idiographic rather than nomothetic methodology.[28] A historical digression to consider Alexander von Humboldt's *Naturgeographie* helps establish this point. At the heart of Humboldt's ideas is the bourgeois subject who perceives and describes 'his landscape'. Aesthetic nature on the one hand and scientific nature on the other become united in a single practical entity, the subject, resulting in an epistemic perspective that makes the metaphysical and the physical level appear as a unity. This perspective and, along with it, the conceptual preference for individual, specific places becomes a central element in constituting, initially, the object of geography and, later on, that of other field sciences as well. Humboldt's methodological physiognomy constitutes a concept of science that offers a 'new form of empirical natural science alongside the experimental one', at the heart of which lies the notion that 'the value of the universal consists in the fact that it provides sympathetic guidance'.[29]

This approach relates to a conceptual distinction identified by Wilhelm Windelband, according to which the experiential sciences are able to operate in the mode of either a nomothetic or an idiographic methodology. The latter focuses on the individuality of a form or a place, conceived of as a manifestation of the universal in the particular, the particular being the measure of validity. When contrasted to a nomothetic methodology, which is guided by the idea of lawfulness, regularity or lawlikeness, this idiographic methodology comes to acquire relevance for virtually every one of the field sciences, whether geography, geology, ecology, ethnology, sociology or hydrology, united as they are by being strongly tied to place. Yet it is not place as an individual unit of measurement that is decisive here, but rather place as an individual form, or gestalt, that can have universal significance.

This connection to place is also what distinguishes the disciplines just mentioned from the laboratory sciences: the latter aim to establish a variety of lawlike knowledge that is independent of place to the extent that the objects of study are either brought into the laboratory or, more usually, come into being there in the first place and are designed as movable objects. These might be, say, lab animals or special plant breeds geared towards a specific experimental design, such as the various types of mice produced at JAX laboratories, or some other techno-phenomenological artefacts such as preserved frog muscles or cell cultures. The 'objective' and nomothetic method is used in the so-called sciences of law (*Gesetzeswissenschaften*), and is frequently identified with the natural and laboratory sciences.[30] The main aim of this method is to generate universal knowledge which is represented best in general statements and mathematical formulae. Here, space – like time – is an abstract and geometrically determined category which, at first sight, appears to manage without any subject at all.

As has already been pointed out, the lab-field boundary is not only a matter of the appreciation of a physical place, its institutional context and its specific practices. Rather, it is also – perhaps to an even greater extent – constituted through epistemological boundary work undertaken prior to the beginning of the 'laboratory revolution' which, according to historians Andrew Cunningham and Perry Williams, dates back to about 1840 and involved the interplay of nomothetic and idiographic methods.[31] In 1828 one of the most famous fieldworkers of the time, Alexander von Humboldt, argued forcefully against 'degrading' the sciences with the collection of empirical data outside the laboratory and the incremental accumulation of specific case studies.

> I have endeavoured, at the same time that I presented the detail of phenomena in different zones, to generalize the idea of respecting them, and to connect them with the great questions in natural philosophy ... These subjects, I believe, are not mere vague theoretical speculations; far from being useless, they lead us to the knowledge of the laws of nature. It would degrade the sciences to make their progress depend solely on the accumulation and study of particular phenomena.[32]

The methodological problem linked to this is how to describe regularities and how to link the description of particular cases with universal laws or lawlikeness.

In Chapters 2 and 3 above we have seen not only that the predominance of the deductive method was accompanied by a concomitant neglect of the inductive method, but above all that this predominance cannot be justified logically or methodologically. This has been described as a matter of balance between different 'styles' of science, and it was argued that since the 1970s this

> balance between the styles of science has swung from a situation in which the deductive and the hypothetical-analogical styles enjoyed cultural supremacy to one dominated by the experimental style and, in particular, by combinations of (or alliances between) various styles of science and the technological style.[33]

This approach of considering the constellation of styles at different moments in time fits quite well with the snapshot approach chosen in this study, even though experimentation is given more weight throughout this analysis, thus foregrounding different constellations of technology and science and of the field and the lab. Deliberating the experiment leads to a focus on the question of what form experiential possibilities assume in any given historical situation (that is, the existing technological, societal and cultural conditions) when a snapshot is taken. And now that we have seen how the notion of experiment necessarily begins to leave the laboratory and becomes further expanded in the context of biology, we will see in the remainder of this study how it extends out even further in the field sciences, especially ecology, until it finally encompasses all of society. All of this points to the fact that any of these constellations involve cultural values relating to the generation of knowledge as well as scientific values regarding the importance of certain kinds of knowledge. This is particularly pertinent and needs to be acknowledged when it comes to judging what kind of knowledge is important and what is not. Just as the interplay between the aesthetic and the universal is rendered specific in the embodied form of the field researcher, this interplay of cultural and scientific values in the field experiment is a hallmark of the idiographic method.

Dirty Places of Experimentation

On the basis of the four key elements identified so far – the appreciation of a physical place, its institutional context, its specific practices and the epistemological boundary work – we can now consider a number of specific practices and examples of field experimentation. As part of my discussion in Chapter 3 on 'When does an Experiment Begin: Experiment and Observation Distinguished', I proposed three criteria for characterizing field experimentation. To pinpoint a scientific object in the field it is necessary to identify, first, what boundaries are relevant and how they should be conceptualized; second, what kinds of sense data are to be collected; and third, how these measurements can be realized in terms of types of instruments and manpower. While place – that is, the idiosyncratic yet representative character of the field – plays a role in all three steps, it is also of utmost importance when it comes to defining the boundaries of the system. The order of the three criteria does not imply the cognitive or epistemic primacy of any one of them. The field experiment might start with any of the three.

The following five cases of experimentation in ecology represent different constellations of field and lab elements. The examples chosen give an idea of the wide range of possibilities for combining the features of the field and the lab ideal.

A certain overlap can be observed with common schemes for characterizing the difference between the natural and the social sciences (or the natural sciences and the humanities), especially in the discourse of interdisciplinarity. The polarities usually offered in this context are human-natural history, human-natural science, historical- nomological explanations, culture-nature.[34] The combinations tell as important a story as the identification of the lab and field ideals itself.

Although the identification and investigation of these ideals tend to reinforce such familiar polarities as those between the hard and soft sciences, between observation and intervention or between nature and culture, it is characteristic of experimentation and the (de)liberation of the experiment that it overcomes these polarities. The impurity and hybridity of experimental practice invites a language of *sowohl-als-auch*, of the *tertium datur*, or 'this-as-well-as-that'. At the same time, the five cases help clarify – systematically if not historically – the key argument of this study of experiment: the liberation of the experiment from the laboratory outwards into the field and further out into society.

The first example is a lab situation set up explicitly to simulate the field – not digitally using computer technology, but using an extracted segment of the 'real' situation (hence the designation real-world simulation). The second example deals with microcosm experiments or, more specifically, the type of mesocosm experiments that are performed outside the lab. These experiments present themselves as the flipside of the first type in that they simulate a lab situation in the field. The third example is about nature itself conducting an experiment – probably the most pure design of a field experiment. The fourth example also concerns an experiment conducted by nature itself and complements the first: it is a real-world simulation consisting of a built object in the field.[35] Finally, we turn to a case where scientific research is itself becoming an agent of experimental change in society.

Tank in the Lab: Enhancing Control

The Max Planck Institute for Limnology in Plön (northern Germany)[36] installed two so-called plankton towers in its laboratory – 'pillars to science' as they were dubbed. The plankton towers were two steel pipes 12m in height and 85cm in diameter, filled with around 10,000 cubic meters of lake water. Every 50cm the column was equipped with a set of sensors for measuring and controlling temperature, pH and light in situ. Each of these water packages is therefore a cross-section of a lake, not unlike the ice-cores that are drilled in the arctic to sample a thick sheet of ice – except that these liquid packages are now isolated from the lake as a very dynamic rather than a frozen system. Interventions in these water packages were done by injecting chemicals or algae and by extracting 'lake water'. In this comprehensively controlled water column, experiments were performed either with particular plankton species or with plankton communities. The whole experimental setting was geared towards better understanding the mechanisms of competition and reproduction of the plankton populations and was thus regarded as fundamental research. The institute's website stated programmatically that the plankton towers were intended to fill the gap between lab experiments and field experiments. At the same time, researchers stated that this was also a source of the problem they faced when attempting to transfer the facts discovered in the towers to the field situation in a lake.

What we learn from this case, which is oriented more towards the lab than towards the field, is that it is precisely the elements of instability and contingency – the dynamics of the lake as a whole – that make it necessary to control the environmental conditions as well as to control the space to perform these experiments. At the same time, these experiments gain their epistemic value precisely because they are related to the field in this way – even if the character of this relationship cannot be fully understood since one does not know just how the experimental system differs from the lake due to the act of isolating the columns. Karin Knorr-Cetina speaks in cases like these of a gap that will never disappear completely, even if the laboratory allows for an improvement (in the sense of better understanding and control) in the relationship between the natural and the social order. One might say that this kind of laboratory experiment constructs and simulates the phenomenon whereas the lake itself, as a pure field experiment, constructs and prompts the emergence of something rather more unruly than a 'phenomenon'.

Mesocosm Experiments: Simulating the Lab Outdoors

The power of the mesocosm experiments is generated by their mediating property and by their capacity to provide local knowledge. The term mesocosm is used to designate an experimental configuration either in scientific or more technologically oriented research, such as ecosystem research or biological monitoring, such as input assessment and pollution control. In a programmatic article published in the 1980s entitled 'The Mesocosm', well-known ecologist Eugene P. Odum noted that '[t]he term *mesocosm* seems appropriate for such middle-sized worlds falling between laboratory *microcosms* and the large, complex, real world *macrocosms*'.[37] Today, mesocosms are widely used and researchers refer more or less explicitly to their 'in-betweenness' and their 'reality push' of sorts.[38] A mesocosm system may occur in quite different shapes, in all kinds of material and, of course, in all kinds of places: they can be located offshore or onshore, they can consist in open or closed containers or basins, they might contain sediments or just water, be charged with macro- or/and microorganisms, with simply one or several testing substances, and they may have the size of a wash tub, a barrel, or a tank and the shape of a pillar or a pond.[39]

It is mainly experimental practice that is responsible for generating the mediating property of this middle-sized world, as experimental practice creates the material-based connection between a local situation outside in the field and a particular, controlled scale and place. Living systems, including organisms, are treated as productive models that are doing a good job by sustaining a certain balance of parts and wholes which can be studied by doing experiments on the mesocosm scale.

The quest for control on a local scale was already present some ninety years ago, when the mesocosm had been invented as a testing facility to do experimental studies on the nutrient conditions and their dynamics in Swedish lakes. Einar Naumann characterized his so-called half barrel method as follows:

> The half barrel method was worked out in order to do experimental studies on the requirements for productivity in different types of water bodies ... In doing so, the concepts eutrophic and oligotrophic were tested experimentally, both of them being of fundamental importance for productivity problems as well as regional limnology in general.[40]

Local mesocosm models make it possible to generate knowledge that acquires momentum towards generalization mainly by applying experimental practices, reducing and controlling parameters and being able to reproduce an experimental setting. Accordingly, mesocosms promise a better understanding of the work going on in the real-world ecosystem. At the same time the mesocosm is itself an artefact, 'a work', insofar as it is an artificially designed object, a more or less skilfully constructed analogical model of the macrocosmic ecosystem.

Nature's Experiments: Invasive Environment and Species

A red and white striped ribbon is all that protects this scientific object – a group of plants, the species *Heracleum mantegazzianum* – from its environment in the grounds of the Center for Interdisciplinary Research (ZIF) in Bielefeld. The milieu of the plants is dominated by wind and precipitation events and is regularly disturbed by curious deer, humans, falling branches and, in some cases, aggressive conservationists waging a struggle against this alien plant species. This 'invasive' species is identified with the problem of biological invasions, which has attracted increasing attention in ecological investigations. The resulting literature is impressive, as is the variety of statements made to explain why this research is important. The following quotation is representative of such statements:

[B]iological invasions represent great natural experiments for the ecologist whose
investigation is extremely valuable for the understanding of population spread and
community- and landscape-level processes affecting the patterns and abundance of
species at large spatial and temporal scales, i.e. scales which are otherwise barely accessible for experimental ecologists.[41]

This strip of meadow with the invasive plants is a virtually uncontrolled place
where (most) instabilities and contingencies are welcomed because they are the
objects of scientific investigation. The special quality of the experiment consists
in the minimal invasiveness of the experimenter, which enables the agency of
nature to come to the fore.

Another type of nature's experiments are natural catastrophes in the past that
have been reported and thus become a source of phenomenal data for recent
climate research. A famous example is the 'year without a summer', the cold
summer of 1816 in Europe, which was due to the eruption of a volcano on the
Indonesian island of Sumbawa. As this event is already in the past, the danger
of direct invasiveness of the experimenter is not an issue. Instead, the question
of bringing nature's agency to the fore becomes a question of the selection and
availability of the historical sources and by the interpretation of the researcher.[42]

Monitoring Nature-as-Experiment: 'Chicken Creek' (Brandenburg)

The object in question is an artificial water catchment. Chicken Creek is a
constructed natural site, a small hill several hectares in size with an altitude
difference of approximately 10m and a small lake at the lowest part of the site.
It is situated in a former strip-mining area in northeastern Germany close to
Cottbus. To avoid the typical problems of such sites (strong soil and water acidi-
fication due to pyrite oxidation), the substrate of the catchment was taken from
sandy accumulations. This sand heap constitutes an ecosystem in its own right.

With the end of the construction work ('point zero' of the ecosystem development)
in the fall of 2005 the first measuring devices were installed, in order to document the
initial state of the system as well as to study the genesis of structures.[43]

The water catchment Chicken Creek is an experimental system that is expected
to enable a closer analysis of the role of physical, chemical and biological parameters and of how they develop and interrelate in different states of the system.
It is an artificial system that was designed to study the characteristics of natural
systems by subjecting it to the same natural processes, presumably, that a natural
system would have undergone millions of years ago when it was *in statu nascendi*.
This system is therefore a specific kind of field experiment that eliminates the
carefully maintained spatial separation between an experimental system and the
natural system, features of which it is supposed to represent. The artificial water

catchment exhibits its own performance parameters and invites the discovery of causal dependencies between different parameters. It does not require an understanding of the way in which it represents 'real' water catchments, since Chicken Creek is quite real enough to substitute for any real system with which it shares dynamic properties. Chicken Creek is, first and foremost, a *real-world simulation*. It becomes a simulation *in silico* only in a second step, when the system is mirrored in a computer model.[44]

Chicken Creek is a technoscientific object in the proper sense: it gathers together theoretical knowledge, instruments, skills and purposes. The artificial water catchment system was designed artificially but is treated as a natural system. It pursues the lab ideal of total experimental control in a field experiment.

Piloting Innovation in Society: Restoring a Lake

Restoration projects can be described in terms of being real-world experiments; they entail a recursive learning process, a typical feature of innovation-oriented projects. Experiments of this type are grounded in the idea that it is not only social and technological innovation that requires scientific experimentation to be extended beyond the laboratory but also that all relevant aspects of society and nature are necessarily involved. The restrictive and protected closed space of the laboratory is left behind. Instead, experimental devices are brought outdoors. Testing stations, prototypes, pilot installations, ecological restoration projects, test releases of drugs, pedagogical reform projects, town district developments and so on all expose scientific knowledge to the unrestricted conditions of unmediated reality. At the same time, projects of this kind require planning, monitoring, data processing and interpretation. The purpose of innovation experiments is 'to turn the relationship between action and surprise into an experimental design'.[45]

Ecotechnology is an example of a disciplinary field that predominantly engages in this kind of experimentation by offering ecological design as a solution to so-called edge problems, that is, conflicts between different agencies, organizations and stakeholder groups. Ecological design is regarded as a technology 'where the form and function of technology assists in improving or, at the very least, only negligibly impacting the environment'.[46] The aim of this type of experimentation is to identify and tame surprises which would never occur in the laboratory world.

In the following, a case study is offered to illustrate this type of experimental intervention in society. The case in question is the restoration of Lake Sempach in Switzerland.[47] A detailed reconstruction of this project has shown how the specific conditions of a lake highlight the need to study its individuality in the context of an experiment in society. Lake Sempach is a typical pre-Alpine lake in the Swiss mid-plains close to the city of Lucerne. The lake's surroundings are densely populated and subject to intensive agricultural use: between the 1950s and the 1980s the population of pigs tripled and many areas of land were drained. The outcome of these activities was a dramatic change in the concentration of phosphorus in the lake, which increased from 25 to about 125 mg/l. Just how dramatic this was in terms of lake pollution became apparent partly due to manifestly adverse phenomena (such as local fish mortality, a high concentration of E. coli and therefore a ban on bathing) and partly due to the changing character of the lake (decrease in the fish population, high production of algae). The qualitative shift was a consequence of the lake having transgressed the threshold for the input of total phosphorus, defined to be 4t/year. This had already happened in the 1960s, with the lake subsequently changing from a so-called mesotrophic to a eutrophic lake.

The scientific problem behind the lacustrian eutrophication problem, namely, identifying the relevant processes and components of the phosphorous cycle, is relatively complex and requires a comprehensive analysis not only of the so-called water column (chemical and biological data) but also of the lake sediment. However, scientists managed to build a model of the lake that allowed them to predict the consequences of any technological intervention. It soon became obvious that the construction of an expensive wastewater treatment system would be a necessary and yet not sufficient component in changing the lake's condition for the better. Things only started moving when all the stakeholders (especially the pig farmers) were included and when several measures were implemented in parallel, not least a reduction in agricultural nutrient run-off and the installation of a ventilation system in the lake. This experimental intervention, however, afforded unexpected results, with the lake reacting completely differently than was expected. Most significantly, the phosphorous concentration did not decrease but increased instead. This surprise finding prompted a research programme on the question of the resolution of phosphorus from the sediment under different conditions, eventually resulting in an enhancement of the phosphorus model: the input from outside the lake is key to triggering sediment resolution in the lake.

It was subsequently necessary to undertake a number of extensive as well as expensive socio-technical measures: the drainage system was reconstructed in its entirety, the effluents were remediated, and the farmers were included in a 'lake contract' aimed at regulating the kinds of crops they grow and the crop sequence, agreeing to abandon the tradition of the winter fallow period. Today, Lake Sempach is once again a mesotrophic lake, the phosphorus threshold of 30mg/l no longer being exceeded.[48]

7 STRETCHING THE BACONIAN CONTRACT –
BUT HOW FAR?

Francis Bacon articulated what came to be called the 'Baconian contract' between modern science and society. The key idea behind this contract is that scientific experimentation cannot cause harm to society as long as it is performed within the walls of scientific institutions. Since the seventeenth century, these walls have served the ideal of an unconstrained quest for facts and the construction of arte-facts for the purpose of expanding and testing the knowledge base of science. The Baconian contract is based on a structural analogy between the methodological isolation of experimental systems from their natural environments and the ideo-logical isolation of experimental activities from their social environments. If the normative structure of society is prepared to permit all kinds of investigations into the causal structure of nature – the nature of human beings and of society alike – then science is prepared to provide all spheres of society with potentially useful knowledge and technology in return. Bacon's lifelong (albeit unsuccess-ful) efforts to gain political support for organizing experimental research on a large scale caused him to ponder the question: what kind of institutional set-ting would convince society of its benefits? Since the promise of gains cannot be justified by an anticipatory form of argument, he suggested that balancing social costs and benefits was a matter of risk and trust:

> For there is no comparison between that which we may lose by not trying and by not succeeding; since by not trying we throw away the chance of an immense good; by not succeeding we only incur the loss of a little human labor ... It appears to me ... that there is hope enough ... not only to make a bold man try [*ad experiendum*], but also to make a sober-minded and wise man believe.[1]

This assessment of the risks associated with political authorization of the experi-mental method was based on an important normative claim concerning the relationship between science and society, namely: experimental failures as well as errors of hypothetical reasoning are acceptable because they affect only the internal discourse of science, not its social environment. Mistakes in the labora-tory can easily be corrected, and society is affected only in terms of its choice of options from those made available by approved, certified scientific knowledge. The isolated laboratory is in a sense a precautionary practice that protects society from experimental failures and errors of hypothetical reasoning. Society is free – and has the responsibility – either to adopt and apply scientific knowledge or else to reject it. These conditions of experimental science have served as the backbone of a dominant ideology that supports scientific progress, making scientific research and technological invention key features of the process of organizing and modernizing society and its institutions. This Baconian conception of experimental science provided a foundational element in the contract between science and society[2] and between society and nature.[3] As the dominant mode of legitimizing scientific research, it prevailed throughout the nineteenth and most of the twentieth centuries. The Baconian contract remains powerful even today.

However, even in late Renaissance society the experimental spirit with its new style of innovative practice spread out into many fields: artists, engineers, instrument makers, surgeons and other practitioners developed new attitudes towards understanding nature's inventions, designing machinery and exploring the globe. Over the course of the mutual development of science and society, experimentation became a polymorphous concept and its social relations multiplied. On the one hand, the ideal of laboratory experimentation became more rigid in terms of reproducibility and precision as well as in its function to serve theory formation. On the other hand, this ideal was modified as it spread to include all kinds of objects – non-living and living, psychological and social, natural and technical, simple and complex, constant and changing, very small and very large, commonplace and unique, within well-defined and ill-defined boundaries, with well-controlled parameters and under uncontrolled field conditions.

The so-called industrial revolution of the nineteenth century has been compared to the 'laboratory revolution' which contributed to it.[4] However, scientific experimentation also became linked with experimental practices of innovation in various sectors of society, such as in agriculture (testing Liebig's artificial fertilizers), in food processing (the bouillon cube, experimentation with calories[5]) and in the health sector (the vaccination campaigns of Pasteur and Koch). Social experiments in urban planning, psychology and policy making likewise contributed to the extension of experimental practices into new fields of inquiry. Even in the natural sciences a shifting of the boundaries became apparent, mainly in biological and geographically oriented disciplines such as oceanography and aquatic ecology: At the beginning of the twentieth century entire lakes were used to perform experiments aimed at finding out more about the lacustrian nutrient cycle. The Schleinsee, a small lake in southern Germany located within a densely populated cultural landscape, became a famous experimental system, enabling some of the most important questions concerning the complex phosphate cycle to be resolved. For this purpose the whole lake was artificially fertilized with phosphates.[6] Today this experiment would be designed to include many more actors, such as conservationists, residents, water sports enthusiasts, fishermen and – if we are prepared to include non-human agents – ducks and water fleas, among others. It would probably be rejected on legal, political and ethical grounds.

Laboratory experimentation is still the dominant approach in knowledge acquisition, particularly in all those research activities and new technologies that imply risks to life or health. However, the attempt to set limits on the spread of theoretical ideas that might entail risks to minds or morals was a hopeless cause from the start. Today, it almost goes without saying that scholars are perfectly at liberty to publicize and even argue forcefully for the application of theories that are still in the making – this holds especially (at least for the most part) in democratic societies in which the principle of freedom of opinion ensures this kind of public deliberation. The premature testing of half-baked theories is an inevitable concomitant of this trend. Beyond this, though, science and technology-based attempts to participate actively in the modernization of society have become more and more successful and have left traces in many arenas. To date legitimacy for these attempts has largely been provided (whether explicitly or implicitly) by the Baconian contract, its elasticity gradually stretched to the point of exhaustion. Thus a new formula capable of handling the interaction between research and innovation is called for – and is already being demonstrated in certain fields.

We are seeing the spread of new styles of experimentation to many areas of society. Experiments performed in open spaces may take the form of a specific kind of social reform or a new law on energy saving, an ecological remediation or compensation measure, or perhaps a technological innovation (such as nuclear power stations or mobile phones). A number of concepts are in circulation that seek to give a label to these various experimental constellations, including 'real-life' or 'real-world experiments', 'experimental installations' or 'experimental innovations', 'social experiments', 'adaptive' or 'experimental management', including 'participatory ecological design',[7] 'living lab' and 'prototyping'.

Experimentation and Innovation in Society

The dynamic interaction between research activities and innovation strategies forms an important feature of what has also come to be known as the 'knowledge society'. One of the most prominent features of this conceptualization is the continuous shifting of knowledge production into contexts of application and a concomitant increase in research in the applied sciences. This trend not only signals the growing relevance of applied knowledge in all domains of society; it also implies the extension of research practices to sites outside the institutional framework of science. And if research – both basic and applied – involves experimentation, experimental activities can increasingly be expected to pervade every field of innovation in society. At the same time, innovation through experimentation is increasingly perceived as inevitable not only in the fields of science and technology but also in the arts, in the private and public world of social relations and activities, and even in the ways in which biographies are composed. In the process, the institutional rationality of science that welcomes errors and failures as a vehicle for augmenting and substantiating knowledge is transferred to society, at least to some degree. Society in turn confronts science with new responsibilities regarding the risks associated with experimentation in the open spaces of societal change.[8]

In the philosophy of science, these changes that are occurring in science and technology in their applied contexts have only become apparent in the recent past, partly through the pivotal role of experimentation and the vast array of scientific practices.[9] Martin Carrier, for example, proposes an 'interactive view' of the relationship between science and technology. He argues that 'for letting this potential of reciprocal stimulation unfold [it is essential] to leave room or leisure for hooking up the practical goals with the theoretical framework'.[10] This evokes a concept of research that comes close to the Baconian notion of the ideal mode of interaction between science and society: maintaining an awareness of epistemic challenges as they emerge alongside societal needs and ensuring that they are reflected in practical research goals. However, Carrier is not the only one to emphasize this interactive relationship. In his book *Pasteur's Quadrant*, first published in 1997, Donald Stokes argues against a linear, unidirectional model of knowledge production and transfer from basic/pure science to applied science and technology. Instead he suggests that technological advances can also lead to a deeper understanding of a particular theory. He puts forward a four-field scheme based on the parameters of 'seeking basic understanding' (understanding) and 'interest in application' (control), resulting in four different research modes, all of which are present in the world of knowledge production (see Figure 7.1). Working through the logical possibilities, he arrives at the following four cases: a yes in understanding and a no in control is 'pure basic research', in Stokes's terminology 'Bohr' research; no understanding, but the capacity for control is 'pure applied research' and is referred to as 'Edison' research; the totally affirmative field, with a yes in both control and understanding, is, interestingly enough, annotated with a question mark – I will come back to this at the beginning of Chapter 10; finally, the fourth case is obviously also the combination favoured by Stokes himself, as it became eponymous for his book: 'Pasteur's Quadrant' is the field characterized by a yes for control and a no for understanding. Pasteur's Quadrant stands for use-inspired basic research, and it is this approach on which the author focuses particularly and which he seeks to advocate. While researchers in this quadrant are fully aware of the potential real-world utility of their work, they never lose sight of their desire to advance scientific understanding as well. Pasteur's Quadrant evokes the Baconian ideal – strongly supported by Stokes's idea of a symbiotic relationship between science and government in the service of human welfare.

		Interest in Application	
		No	Yes
Seeking Basic Understanding	Yes	'Bohr' pure basic research	'Pasteur' use-inspired basic research
	No	?	'Edison' pure applied research

Figure 7.1: Model of different types of research depicted as a four-field schema; adapted from D. Stokes, *Pasteur's Quadrant: Basic Science and Technological Innovation* (Washington, DC: Brookings Institution Press, 1997).

However, the Baconian contract – the ideal of relating science to society – is not capable of addressing the spread of scientific practices into innovative fields of society. Instead, it is likely that we are heading towards a new knowledge-society contract, which on the one hand turns society into a research field and open laboratory and on the other binds experimental practices and hypothetical reasoning to the conditionality of social acceptance. This shift to a new, innovation-driven constellation between science, technology and society is also reflected in the concept of the knowledge society. Accordingly, the new contract might be called the knowledge-society contract or – perhaps even better, with its reference to practices and material-based processes of innovation – the experiment-driven-society contract.

Social Experimentation: Field Experimentation in Society

As we have seen above, many features of field experimentation are not new, and it need hardly be pointed out that field experimentation as a methodology extends across many disciplinary fields and domains of knowledge. Field experiments are performed not only in biology but also in archaeology, geology, agronomy, ecology and other field sciences. 'Social experimentation' focuses on fields where humans are regarded as the main actors/objects of a scientific study and where it is a matter not only of sociological but also of political, economic or educational experimentation. Equally, however, 'social experimentation' alludes to social relationships as opposed to non-human relationships as research objects and, as such, may also be involved to a greater or lesser extent in experiments in, say, restoration or landscape ecology.

Accordingly, humans may be involved in a more or less explicit way. Their role may be negligible, for instance, in ethological experimentation outdoors or in this or that experimental design in ecology. At the other end of the scale, though, there have been experiments *with* human subjects in urban planning or restoration ecology, for example, which – more often than not – have turned out to be experiments *on* human subjects. This is certainly a problematic phenomenon; it is the dark side of experimentation and of course holds not only for field experimentation but also for experiments conducted in the lab. Certain types of experiment have been particularly contested, vivisection in physiology and medicine being one type,[11] and laboratory experiments with humans along with experimentation in society-as-laboratory being two others.[12] History abounds with examples from medical and psychological research, weapons testing, schooling and learning, colonialism and the chemical industry (to name just a few) where human beings have been improperly exposed to strategies of trial and error.

To be sure, this kind of 'science in the making' has been done in authoritarian and democratic societies alike: in both settings, it is only in a few cases that the research subjects involved and the general public at large have been properly informed (let alone invited to participate actively). Often, the experimental design was both secretive and sloppy. Given this backdrop, this kind of experimentation stands on ethically and politically slippery ground. If it is true, however, that scientific research is increasingly becoming an agent of change, then new forms of legitimacy, information dissemination and participation are needed that can be readily interpreted as giving rise to the emerging contract between science, technology and society. In the following, a few snapshots are offered of the history of experimental interventions in the field of 'society', often referred to as social experimentation or social experiment.

Experimentation as Problem Solving: John Dewey

In the early years of the twentieth century, American philosopher John Dewey was a prominent and influential proponent of the idea that experimental knowledge production and social change are interwoven: 'The ultimate objects of science', he wrote, 'are guided processes of change', and truths are 'processes of change so directed that they achieve an intended consummation'.[13] Certainty in knowledge follows from achieving reliability in action. Dewey refers explicitly to the benefits and values of experimentation, identifying this scientific methodology not only with modernity but above all – and this is the key issue here – with the ideal of attaining a more egalitarian society. Accordingly, it is not sufficient merely to publish one's theory or observations; rather, they must be based first and foremost on a shared rationality (within the scientific community) and on a well-documented practical procedure (in general experimentation) that connects theory and observations. Most importantly, Dewey believes that this strategy for problem solving is likewise applicable to ethical problems and, furthermore, he is ready to melt the boundary between the famous two cultures of the sciences and the humanities on the basis of an experimental mode of problem solving. On the basis of Dewey's pragmatist approach, experimentation becomes a general methodology that enables us to treat ideas as instruments for solving problems (which are always situational and context specific), the mode of discovering involves intervention in the environment (and results in a re-adaptation of organism to environment) and, finally, problem solving is a constantly evolving process to which there is no final solution.

Accordingly, for Dewey there is no difference in kind between the physical world and human nature, that is, between the material and the mental/moral world. In both cases there is no need to reject past experiences (and enjoyment); but these past experiences are not necessarily authoritative for the present or even the future (as can be learned from Russell's inductive chicken). At the same time, we need not imagine that this break leads us towards chaos (as Dewey never ceases to reassure his readers) because we still have customs and institutions, education and a constitution to help equilibrating this kind of change, that is, this kind of social and technological innovation. On the basis of this pragmatist approach, 'doing experiments' means probing facts, values and norms according to changing environmental conditions – and thereby creating a more desirable environment.

Piecemeal Social Engineering: Karl Popper

After World War II, Karl Popper – struggling with totalitarian political experiments – suggested 'piecemeal social engineering' as a way to introduce the scientific method into politics.[14] For Popper, piecemeal engineering is the opposite of what he called utopian engineering, in that it involves a mode of social engineering that does not seek to attain a perfect blueprint of the state but is instead fully aware of the open-endedness of social and political processes. 'The piecemeal engineer will, accordingly, adopt the method of searching for, and fighting against, the greatest and most urgent evil of society, rather than searching for, and fighting for, its greatest ultimate good'.[15] Popper proposes the step-by-step method as a reasonable one that makes it possible to improve not only society but also the individual, since it can be applied at any given moment.

It is contrasted to a method which, although it follows an abstract and ambitious ideal, may easily lead to greater suffering among people because it tends towards 'continuously postponing action until a later date, when conditions are more favourable'.[16] It is quite obvious what kind of horribly misguided political ideals Popper had in mind when he advocated the pursuit of piecemeal engineering rather than social engineering on a grand scale. Developing those piecemeal engineering blueprints, such as educational reform or a different set of health care institutions, is 'comparatively simple', and if they go wrong, 'the damage is not very great, and a re-adjustment not very difficult'.[17] Popper expands his piecemeal argument beyond the mere experimental setting of the physical world and addresses the actors in this world who do better when they seek compromises. It is this moral moderation which, according to Popper, leads to social improvement via democratic methods as opposed to the obsession with improving society as a whole while solely relying on the 'centralized rule of a few'. Consequently he concludes that it is 'making social experiments which alone can furnish us with the practical experience needed'[18] and that we should conceive of modest social experiments only.

'The' Chicago School: Discovering the Urban Field

The notion of societal experimentation developed by the Chicago school of sociology served to radicalize the pragmatist approach as it was put forward by Dewey. In its formative years, it was Albion Small who was identified with this school. He believed that the rapid change occurring in modern settlements provides in itself a 'world of experimentation open to the observation of social science. The radical difference is that the laboratory scientists can arrange their own experiments while we social scientists for the most part have our experiments arranged for us'.[19] Small located the idea of experimentation in social life and not in the scientific method; at the time, the notion of experimentation had become quite influential in American sociology. However, during the heyday of the Chicago school in the 1920s, it was Robert E. Park and Ernest W. Burgess who became the central figures. Both of them can be identified with an ecological approach towards organizational change, studying processes of organization-environment relations. Thomas Gieryn has pointed out that 'the substantive domain of "the" Chicago School depends upon which history or reminiscence you read'.[20] However, there is some evidence that the Chicago school of urban studies advanced a shared intellectual identity, without ever providing a methodologically or theoretically homogeneous universe. Rather, it promoted an awareness of the situatedness of all social processes, that is, the contextual location of social facts in space and time.

In 1963 Donald Campbell wrote of experimental and quasi-experimental design, while in his influential paper of 1969 'Reforms as Experiments'[21] he proposed a methodology comprising political planning and the scientific design of social experiments. This type of approach also began to appear in the context of US policy in the 1960s, when several social experiments had already been performed. These included the Manhattan bail bond experiment, motivated by a desire to relieve the pressure on city jails from people awaiting trial. Another was the negative income tax experiment, consisting of four campaigns designed to test the hypothesis that a single guaranteed welfare payment for all poor families constitutes an efficient and effective way of replacing the complex set of cash and in-kind benefits paid previously to certain categories of poor people only. These experimentally designed trials of US government intervention were the precursors of a corresponding, more systematic research programme which finally took off in 1971. In this programme, the Social Science Research Council appointed a committee to deliberate the various aspects of social experimentation (such as design and measurement) along with its practical aspects with respect to sponsorship and the utilization of results. The US National Science Foundation provided financial support for workshops in which leading sociologists such as Donald Campbell and Albert Rees (both members of the Chicago school) as well as Henry Riecken and Robert Boruch participated, both of the latter becoming the editors of a topical book at the time entitled *Social Experimentation: A Method for Planning and Evaluating Social Intervention*, published in 1974.

Although criticisms have been expressed regarding the rather technocratic attitude of the Chicago school approaches (i.e. that reforms were imposed more or less coercively on people, following a principle of downstream policy making), they have been influential in more recent attempts in which those affected by the policies in question have been turned into participant observers.

Altogether, these various new approaches and concepts can, with hindsight, be seen as foreshadowing the search for a new contract between science and society, without, of course, suggesting that there is a coherent line of development or an epistemologically or politically straightforward story to be told. However, the cases and individuals described above – along with the many others not mentioned here – might be useful witnesses whose intellectual and practical work can be seen to have contributed towards the process of advancing a changing mode of experimentation, one that has expanding within and continues to meander through more and more diverse domains of society.

Innovation Necessarily Involves Surprises

The importance of science in society derives not only from its ever increasing contribution towards knowledge but is based above all on the transfer of experimental practices to the design, monitoring and evaluation of innovation processes. The search for new sources of energy is a good example. The development of potential scenarios is unavoidably based on assumptions regarding energy resources, new technologies and consumer lifestyles that are in turn heavily dependent on recent scientific knowledge or even hypothetical reasoning. Thus decisions based on scenarios necessarily contain experimental elements. Monitoring and feedback mechanisms determine the development of novel strategies which are either reinforced by success or weakened by surprise. Even if there is an element of path-dependent lock-in that arises from heavy investment in, say, nuclear power plants, coal strip mining or offshore wind parks, the respective economic, ecological and political cost-benefit ratio informs future decisions. The formation of new political institutions under the impact of experimental research strategies is still in its initial stage. It seems that politics itself is changing as it acquires a more experimental style which – with respect to the European Union – is variously called 'experimentalist governance', 'regulatory experimentalism' or 'collective experimentation'.[22] These moves to embrace new innovation strategies indicate the revision of the Baconian contract towards an again renovated Baconian contract, which may bear in the end an 'experiment-driven society'.

In the search for such a contract, it is important to be aware of the fragile and rather limited possibilities of a fundamental cognitive shift and therefore to advance a more modest view. The opportunities and politics of experimental practices as put forward in this book provide such a modest position. They encompass an awareness of the conflict between the steadily growing number of more or less explicit field experimentation projects on the one hand and the questionable claims to social acceptance and institutional responsibility on the other. Thus the experiment-driven society finds itself in a paradoxical state: the more it absorbs scientific knowledge by pursuing the various available modes of experimentation, the more it is compelled to deal with non-knowledge and its variants ambiguity, ambivalence and indeterminacy.

If experimentation is indeed a privileged way out of this dilemma, then the institutional tools of policy making need to be further developed. This is where a moderate stance comes in – one that does not overestimate what a renewed Baconian contract can do in terms of forging a consensus. Instead, it is a matter of *learning* to soundly manage the paradoxical state of knowledge and the tension between the commitments of politics and of experimentation: politics is bound to the rhetoric of right and wrong, while public opinion is strongly oriented towards preventing and avoiding risks. An experimental attitude, by contrast, requires hypothetical reasoning and runs the risk of failure. Unlike the concerns voiced by the advocates of the risk society (where failures are seen mainly as malfunctions of the system), an experiment-driven society conceives of failures as constructive components of learning. Certainly, there are cases where this kind of learning does not seem to be very systematic or where it is misinterpreted – as, for instance, when society and its members are viewed rather like a population of rats, as in the Milgram experiments.[23] A further, cynical example would be anthropogenic climate change, had it been planned as a global experiment.

Today, there are legal norms in place that set limits to experimenting with uninformed people. The precautionary principle, codes of conduct and experimental protocols are all intended to prevent the burden of risks being shifted onto future generations. Most innovative fields of technology such as nano-, eco- or biotechnologies, man-machine communication, medical research, energy issues and conservation merge basic research and experimental application with unknown consequences. Consequently, this is the arena where an institutional and conceptual framework is required to provide a proper conceptualization of extended field experimentation consisting of epistemic norms, political guidelines and policy procedures. Experimental governance would include making the experimental design explicit, establishing procedures for monitoring and data processing, and devising mechanisms for keeping the public informed and negotiating with concerned citizens and interest groups or, even better, enabling them to participate.

Whether or not such requirements slow down or accelerate innovation strategies is hard to predict. In the days of classical Baconian science it appeared that the Baconian contract not only gave science its indispensable, independent and uncontrolled space but also fulfilled the expectations of the contractual parties with respect to the added value of the 'fruits of knowledge' – a perspective that all too often created problems rather than solving them. In our own times a different era holds sway, one that is defined, first, by the still-precarious acceptance of science as an agent of societal change, turning much of science into field experimentation, and, second, by growing demands for public engagement to shape and control knowledge production in order to turn science and technology into a democratic endeavour.[24] Thus if there is a new contract between science and society, it is being forged in an experimental mode.

8 ABOUT THE EPISTEMOLOGY AND CULTURE OF BORDERS/BOUNDARIES

The preceding discussion of the renewal of the Baconian contract and the emergence of an experiment-driven society implies a transgression of boundaries in several respects. As mentioned above, when science acts as an agent of societal change, much of it turns into field experimentation and thus goes beyond the boundaries of laboratory research. More than this, though, the growing demand for public engagement to shape and control this scientifically inspired knowledge production affords the requirement to turn science and technology into a democratic endeavour – at least if participatory principles are part of the society's shared system of values. However, there are also other modes of transgression involved which refer to longer-standing areas of conflict, mainly within the domain of science itself. 'Transgressing the boundaries' was the title of a paper smuggled by physicist Alan Sokal – Trojan horse-like – past the peer review system into the well-respected journal *Social Text*. The paper was a parody of social constructivism and served to intensify an already ongoing controversy between scientists and social scientists that came to be called the 'science wars'.[1] Here, the notion of 'transgressing the boundaries' was reduced to absurdity; Sokal's article served to debunk the apparent ignorance of science critics, their intellectual laziness and lack of sincerity regarding their objects, science and technology.

Since then, the science wars of the 1990s have faded away even as technology assessment (*Technikfolgenabschätzung*) and so-called ELSA (ethical-legal-social aspects) research, or *Begleitforschung*, have become well established[2] and have developed their own methodological toolbox and institutional settings. There is general agreement that the transgression of conceptual or disciplinary boundaries is not only a standard situation in research and development, but that the systematic and institutional challenges posed by these transgressions influence and even structure deliberation processes relating to innovation and governance. This becomes particularly relevant when a novel technology is introduced in society or when an existing technology develops surprise – in other words, whenever innovation processes and risk knowledge diverge. The fact that today's STS and ELSA research are seen to play a decisive role in influencing and indeed even shaping the public perception of novel products and technologies suggests yet another phase or cultural moment in the relationship between the innovative research of the natural/engineering sciences and the so-called reflective sciences (*Reflexionswissenschaften*). This relationship appears to be defined by complementarity rather than by reflection following upon and reacting to scientific and technological developments. This idea of complementarity, if not fusion, is underscored when someone like Christos Tokamanis, head of the unit 'Nano and Converging Sciences and Technologies' at the European Commission, proclaims at a workshop on funding directions: 'Nanotechnology is not R&D as we know it but a socio-political project'.[3]

This call for complementary research – another blurring of boundaries – is motivated by a sense that the governance of emerging technologies amounts to acting in real time under conditions of ignorance. At the same time, the demands of knowledge and the scope of concern have been extended to include an indefinite future and the preservation of natural resources as well as the well-being of animals and plants.[4] Such an injunction to anticipate and prevent even unknown hazards over the long term provides the setting in which precaution becomes a normative principle. The state is assigned responsibility for the differential weighting of acceptable ignorance, necessary knowledge production (e.g. the call to falsify concrete risk hypotheses) and the degree of regulation. All this calls for political agency and decision making, which in turn depends on the socio-cultural context: society's willingness to take on risk and its desire to avoid risk, including the transgression of cultural habits and the willingness to enter into collective experimentation in order to manage ignorance. Explicitly and implicitly, all this is deliberated and defined by society.

Focusing on Borders and Boundaries

Over the last few decades, virtually every dualistic conception used to describe scientific knowledge or cultural configurations has come under critique – subject and object, fact and value, nature and culture, female and male, theory and practice, to mention just the most prominent ones. The same holds for science and its institutional setting when one considers the dualism of basic and applied research, of science and technology, of public or academic and private or entrepreneurial research. In the preceding sections it was suggested that this dissolution of dichotomies owes to the growing importance of field experimentation in society and, ultimately, to a society that perceives itself as engaged in experimentation. Correspondingly, it can be argued that the focus on borders and boundaries and on their maintenance and transgression is advanced by this expansion of the experimental mode of tinkering – of trial and error – in society.

Accordingly, two propositions lay the ground for the following analysis: first, boundaries are coming to the fore because they appear to be vanishing, along with them the dualistic pattern that has been taken for granted in modernity. Second, conceptions of border or boundary are positivised in the many accounts of 'the inter-disciplines',[5] of transdisciplinary research and its boundary objects, its boundary work and boundary concepts. The following reflections on borders and boundaries seek to explore the images and imaginations associated with this increased atten-tiveness and declarative process of putting a positive emphasis on 'bordering'. They require a terminological clarification which is itself part of the story.

Grenzen (Limits, Boundaries) and *Schranken* (Borders)

Philosophers frequently point to Immanuel Kant's canonical distinction between constitutive *Grenzen* (limits, boundaries) and merely empirical *Schranken* (bor-ders) in order to maintain a classical dualism between the spheres of rationality, necessity and value on the one hand and of reality and empirical contingency on the other.[6] Accordingly, the suggestion was made in 1972 at the first UN confer-ence on the human environment in Stockholm to use the term 'boundary' to denote the normative bounds of action (Kant's *Grenzen*) and 'border' to describe physical limits (Kant's *Schranken*). The recommendation did not catch on, how-ever, and in subsequent years the terms border and boundary and their normative and descriptive elements have become conflated. To reflect this hybridization, in the remainder of this chapter I will adopt the convention of writing border/boundary. Moreover – and again, one can refer to Kant as an example – the dis-tinction between border and boundary involves a kind of visual imagination: the hybridizing movement that takes us beyond these concepts involves the visual metaphors of blurring boundaries and transgressing borders. Both of these will play a role in the following discussion of border/boundary in environmental dis-course. The main backdrop to this discussion is the 'limits to growth' debate and the 'border traumas' of the 1960s and 1970s. The issue came especially to promi-nence in such well-known publications as Rachel Carson's *Silent Spring* (1962) and *The Limits to Growth* (1972) edited by the Club of Rome.

The 'global existential crisis' of the second half of the twentieth century made environmental and ecological issues matters of public concern. Initially at least, the 'limits to growth' were conceived above all as ontological borders/boundaries which are believed to be simply given and non-negotiable. This onto-logical interpretation was gradually overshadowed and eventually replaced by an epistemological one according to which the limits are relative to a specific state of knowledge, social organization and technology. This gave rise on the one hand to the discourse of sustainable development, sustainability studies and the environmental sciences. At the same time, it led to the emergence of an environmental discourse on disciplinarity and to a discourse of interdisciplinarity within ecology and the environmental sciences. Indeed, the distinction between multi-, inter- and transdisciplinarity can be mapped onto the development of ecology and the environmental sciences. 'Multidisciplinarity' generally refers to a variety of perspectives on a given problem or question, whereas the 'interdisciplines' are constituted by a highly complex problem that owes its complexity to its origin in the real world and to the societal expectations or needs associated with it. Finally, the activities pursued within transdisciplinary (or also so-called mode 2 research) seek explicitly to transgress academic borders. These activities exhibit a certain agency, serving as a driving force that pervades conceptions of the knowledge society. A number of conceptions have additionally emerged that refer directly to the border/boundary, such as the border zone and the trading zone, along with boundary work and the boundary object.[7] It is these distinct 'experimental practices at the border' which will be scrutinized in the following, along with a focus on the interdiscipline 'ecological knowledge and practices'.

Socio-political and Scientific Ecology

The nature of ecological knowledge and its significance for socio-political deliberation in the shaping of modern societies is widely accepted nowadays: '[I]n the last 50 years our industrial societies have turned out to be ecologically adaptable and they will continue to be so in the next 50 years', asserts sociologist Wolfgang van den Daele,[8] while ethicist Konrad Ott subscribes to a 'departure and change – regimes for a new green deal',[9] which is the title of a report in which he focuses mainly on a critique of neoliberal economics and its stakeholders whom he sees as hindering the development and deliberation of ambitions and policies that can and must be pursued by civil society alone.

The idea that boundaries are an important – and even essential – issue within ecology is once more supported by, among others, urban ecologist Herbert Sukopp in his book *Recapture? Nature in the City*. Ecology in general and the study of urban habitats in particular

> picks up on social problems and participates in the process of defining problems, their causes and appropriate strategies for solving them. It relies on interdisciplinary collaboration. In addition to describing an actual state, it also calls for judgments and the formulation of target conditions ... In this way the borders between science and practice, between disciplines as well as between value-free, 'objective' science and planning, become blurred.[10]

These reflections on the disciplinary border of ecology and its blurred character are in a sense historically in-built attributes of ecology that are also reflected in the history of the concept itself. From the late nineteenth and early twentieth centuries onwards, 'ecology' was variously described as 'ethology' or 'biology', while such divergent terms as 'physiological' and 'sociological' were used to demarcate the term 'ecological'. The first half of the twentieth century saw a rapid blossoming of scientific societies, academic institutions and publication outlets so that by the end of this period ecology came to be described as a 'super-science', first in relation to the rise of systems theory and particularly of cybernetics in the 1940s and then again with the arrival of the environmental movement in the 1960s, as an answer to the ecological crisis. In this broadened sense, 'ecology' served (more or less explicitly) to blur the boundaries between scientific, philosophical and political knowledge. At the methodological level this involved a merging of facts and values, of the epistemic and the social. These blurred and hybrid conditions became the subject of heated debate and triggered a period of frantic boundary work between scientific ecology and other fields such as cultural ecology or political ecology. From this time on, the term 'ecology' could be used to refer to a variety of completely disparate scientific approaches, ideological doctrines and political stances. The struggle to define 'ecology' has therefore always been one of both interpreting the complex subject matter of 'the interdependence of living and non-living nature' and debating the relationship between institutional and social groups in academia and beyond. This process is still ongoing and may well prove to be of crucial importance in the context of twenty-first century environmental matters.

Given that the production of ecological knowledge – and even more so that of the environmental sciences – has to be seen as being constitutionally located at the intersection of various borders/boundaries, the knowledge thus produced and labelled as ecological will be crucially influenced by the particular character of those boundaries. This tenuous position renders it experimental in that every hypothesis proves to be an intervention with unpredictable repercussions for the various fields delineated by the intersecting boundaries. To use a Foucauldian expression, we might speak of ecological knowledge as being structured by a '*dispositif* of borders'. This *dispositif of borders* provides an experimental precondition as well as a structure for ecological knowledge. It also constitutes the theory and practice of ecological action and structures our way of looking at societal discourses on nature. Perhaps we can start here to formulate a general framework for a *dispositif of borders* by investigating and comparing some of the key concepts that have been used to represent how knowledge domains interact and coalesce.

The production of ecological knowledge implies the construction of boundaries; in addition, this procedure is necessarily linked to a particular place, namely, one that is either virtual (in the case of concepts or models) or real (involving physical entities). It is appropriate, then, to focus on how boundaries are drawn and in what ways knowledge is framed by their invention, constitution and normalization, and by their formation and arrangement. Building on boundaries means incorporating knowledge and thereby transforming it. Conversely, the contours and qualities of disciplinary boundaries are influenced by the emergence of new fields of knowledge. Thus the following hypothesis suggests itself: the imagination and conceptualization of borders/boundarys is intertwined with the imagery of borders/boundaries. In other words, the conception of a border/boundary will depend to a certain extent on the visual representations made of borders/boundaries.

Epistemic and Epistemological Boundaries

These considerations suggest that it is important to look at the use of language, that is, at the language games in which border/boundary are deployed in this reciprocal process of borderland constitution and knowledge formation. The extension and intention of the boundary concept, its metaphorical use, its practical application, and the kind of description applied to it all determine equally the shape and arrangement of the boundary situation and therefore of knowledge itself. This view of language and its relation to forms of life and conditions of knowledge draws on Wittgenstein's *Philosophical Investigations*. Significantly, in advancing this view, Wittgenstein adopts the experimental mode of the aphorism, thereby inviting multiple interpretations of his statements. For example, there are different ways of reading the remark 'Philosophy is a battle against the bewitchment of our intelligence by means of language'.[11] In an experimental manner, this sentence can be configured either as asserting that philosophers use language to combat bewitchment or as asserting that philosophers combat a bewitchment that is caused by language itself – and, of course, it might be taken to mean both. As with the famous duck-rabbit and the problem of aspect-seeing, the perception of a particular gestalt depends on context and boundary conditions. The point Wittgenstein wishes to make here is the ambiguity of sentences and of figures that cannot be resolved. Wittgenstein stated quite openly that he was not exactly sure what is going on when this kind of ambiguity appears. What he was sure about, though, was that it wasn't simply the case that the external world stayed the same and an 'internal' cognitive change took place. For Wittgenstein, meaning, thought – and therefore, one might add, knowledge too – is ineluctably social: using language is an activity. In a book on Wittgenstein's *Tractatus* this was explained as follows: 'When I say that I love you, Angela, I am reaching across an unbridgeable abyss ... the meaning of the sentence of "I love you" lies in the mode of its verification, in our life together'.[12] It is the performative activity of a speech act in the social world that verifies the meaning of language, and only this activity may, if not clarify, then at least temper its ambiguity.

There is another ambiguity involved in the borders/boundaries discourse in that they are always part of at least two different entities. Borders link that which is basically different and are therefore always related to the respective entities they are bringing together. This is the concept of a border that can be also found in Kant, who illustrated this using the following example:

> Area is the border of the [three-dimensional] physical space, while itself being a space; a line is a space that is the border of an area; a point is the border of a line, but still remains a location in space.[13]

He adds that borders 'merely contain negations' whereas there is always 'also something positive' in a boundary (*Grenze*) in that it brings two entities into relation with one another. This 'something positive' of relatedness as well as the 'something negative' that elides the greater dimensions to which it testifies, inform the concepts of borders/boundaries in environmental discourse which are hybrids of normativity and descriptiveness.[14]

Four Perspectives on Border/Boundary

Borders/boundaries can be looked at from different perspectives, with very different consequences for the way border/boundary is conceptualized. An ontological perspective will aim to produce an explanatory statement about the real world, the 'limits to growth' debate being an example of this kind of conception. It nicely demonstrates that an ontological statement might persist, even if the existence of the limit cannot be scrutinized empirically. From the perspective of realism (whether one conceives it as entity realism or as structural realism), limits to growth or ecosystems do exist and they can be perceived by means of delineating their boundaries, even if these boundaries and their delineations can be inferred only from models.[15]

The epistemological perspective is concerned with the limits of models and the limits of knowledge. At this point, conventions, construction and representation become merged: a computer model can show a border where in the real world there is none. This might be accepted for pragmatic reasons, for example, because the model is easy to manipulate; at any rate, it is not taken as grounds for scepticism. On the contrary, limits are seen as a necessary condition of the possibility of distinguishing objects from their environments; accordingly, the conceptualization of a border/boundary implies the constitution of the object. This object, however, need not be fixed but can depend on a state of knowledge. As opposed to the Club of Rome's naturally fixed limits to growth, the Brundlandt report's conception of sustainability maintains that the limits of nature depend on the state of technology and its capacity to exploit nature more or less efficiently.

From a methodological perspective, borders/boundaries constitute appropriate methods and rules for gaining knowledge; their use is merely pragmatic and they are regarded accordingly as cognitive and material instruments. These instruments are simultaneously medium and interface, and they are subject to scientific, sociological and philosophical considerations, for example regarding metrological practices or issues of experimentation and instrumentalism (see also the discussion in Chapter 4 on the material realization of experiments).

Finally, the operational perspective is closely related to the methodological perspective but goes beyond the latter insofar as it incorporates a more complex set of values, normative decision-making techniques and normalizing procedures. Deliberations on threshold values are paradigmatic here of the discourse on borders/boundaries. Threshold values are not given by a physical theory (such as the absolute zero of temperature) and are therefore subject to normalizing procedures and normative decision making. The deliberation of thresholds refers to an empirically measured point at which some degree of harm has been deemed to occur to the environment or to human beings. These are empirical norms that serve as the basis for political action and, ultimately, for legal regulation and regulatory intervention.

These four uses of borders/boundaries do more than just signify different language games or constitute different domains of objects relative to forms of knowledge. They also elicit different ways of probing the boundaries or testing the limits, and thus establish different modes of experimentation. Viewed from the ontological perspective, experimentation explores the conditions of catastrophe that come with the overstepping of limits. From the epistemological perspective, experimentation builds comparative scenarios for staying just within variable limits. From the methodological perspective, experiments consist in fixing limits of sensitivity in order to obtain data at various grades of resolution or coarseness. And from the operational perspective, setting limits is an experimental intervention that tests not just reality but also society and its risk tolerance, for example.

Each of the four perspectives is relevant to the identification of borders/boundaries in ecology and the environmental sciences. The debate about borders within and in relation to ecology obviously also includes debates concerning the role and importance of ecological knowledge in social processes and in negotiations about the kind of environment in which we wish to live. Borders and boundaries are thus omnipresent in ecology and even more so in the environmental sciences. They are considered in the conceptual framework of these disciplinary fields and in the constitution of their objects, and they can themselves become a matter of interest, such as in research programmes investigating biodiversity, the succession of plants, the ecotone or ecosystem services. All these conceptions entail boundaries/borders of different kinds: the physical border, for instance, between water and land or forest and meadow (ecotone, succession), the distinction between pure and applied science (ecosystem services, biodiversity), the contrast between facts and values (ecosystem services, biodiversity, succession), the question of what is natural and what has been culturally transformed (ecotone, succession, biodiversity). This constant involvement in border/boundary discourses is also a permanent challenge for environmental theorists, whether they be philosophers, historians or sociologists. They, too, are constantly confronted by a plethora of diverse borders and boundaries, thresholds and limits: can nature and culture be divided? Are natural divisions discovered or created? How do political borders and moral economies influence community building and help to transform cultural landscapes?

Visual Representations of Borders/Boundaries: A Morphological Study

Laboratory studies in particular have conjured images of overlapping, blurred, displaced and shifting boundaries, of merging, fusing and intersecting domains, and of mingling and migrating individuals. Here, the extension of the experimental mode to the probing of borders/boundaries in ecology and the environmental sciences becomes generalized to comprehend scientific practice in relation to its social environment. If one now wishes to analyse these different semantics and images of the borders/boundaries discourse, one might arrive at a morphology of border/boundary concepts. Since images – including visual images, metaphors and analogies – are widely used to discuss the negotiation of borders/boundaries in the sciences, this imagery will be related to political models of boundary phenomena. This comparison of images or models of borders/boundaries in the contexts of science and politics will reveal more than just 'family resemblances of border phenomena';[16] it will also provide insights into the cognitive structure of knowledge building at the border/boundary. Boundaries thus emerge (in this context) not merely as 'negative' lines of division but as an open, 'positive' arena for co-construction and social learning. As meanings, positions, semantic fields, individuals and objects, real and virtual entities or processes are brought together, the power of boundaries proves to lie in their capacity to create unexpected connections and in their invitation to engage in experimentation.

But why should we be talking here about the morphology of borders? Morphology is the theory of forms. It is a sub-discipline of many branches of scholarship, including biology and linguistics, and it constructs its own disciplines, such as urban morphology and geomorphology. Accordingly, when asking about the form and structure of borders and boundaries, one might also ask what properties they can possess. To offer a few examples, borders can be open or closed, active or passive, sharp or blurred, dynamic or static, coherent or fragmentary, stable or unstable. At least some of these descriptors have been applied to various borderland models in geography and politics.[17]

Boundary Work and Boundary Object, Trading Zone and Border Zone

The following discussion of borders is restricted to the border concepts that have been developed over the last twenty-five years or so within science and technology studies (STS). These are, most prominently, boundary work,[18] boundary objects,[19] trading zone[20] and border zone.[21] All these concepts rely heavily on analogies to borders and boundaries that have been described in the context of political geography and topography.

Gieryn's extensively developed concept of boundary work relates to the scientific practices that construct the social boundaries separating science from everything else (engineering and religion in particular). Boundary work, in this sense, 'is a strategic practical action' performed to monopolize, expand and protect science.[22] Boundary work is first and foremost rhetorical and ideological work carried out in order to compete against other cultural activities. Gieryn discusses this, for example, in respect, to the boundary between physics and technology in the nineteenth century. This boundary work resulted in the positioning of engineering as applied science in contrast to physics as theoretical, 'proper' science – a structuring of the field that has remained in place (and effective) to this day.

Boundary objects, as defined by Star and Griesemer, 'inhabit several intersecting social worlds' and are able to 'satisfy the informational requirements of each of them'.[23] Boundary objects are things whose meanings are sufficiently pliable that people from different social worlds can view them as embodying a shared cause. The concept applies especially well to traditional scientific disciplines, each with their own internal standards, domains of objects and cohesive practices. Certain objects in the research environment help people from these different disciplinary worlds to come together to conduct their own research and make progress on their own scientific problems. The notion of the boundary object brings to light cooperative activities that enable distinct social worlds to achieve their shared objectives in spite of the boundaries that separate them.

We encountered Peter Galison's 'trading zone' in Chapter 4 above when *discussing* the material realization of experiments. The trading zone is 'an arena in which radically different activities (can) ... be locally, but not globally, coordinated'.[24] Key to this conceptualization of the boundary is the development of a new language, beginning with a rudimentary hybrid, a pidgin language, which enables initial understanding. As time passes, the pidgin becomes more differentiated and evolves ever further. In the end, the trading zone becomes a novel, though not disciplinary, scientific community with its own culture, practices and language (creole).

Robert Kohler suggested the concept of the 'border zone' to describe the development of early ecology in North America. Kohler prefers the term 'border' to 'boundary', mainly because the latter reminds him of 'political or property maps, where sharp definition matters (orange flags in the woods, Checkpoint Charley)'.[25] Kohler focuses on the zone between the two different scientific cultures of the laboratory and field sciences, identifying the differences in their methods and practices, the choice and constitution of their objects, the control and recruitment of their actors, as well as the specificity of place in each. Border zones are undefined meeting grounds with mixed practices and ambiguous identities. Ecology – or 'border biology' as Kohler calls it – never gets out of the border zone but always stays in it, resulting in a patchwork discipline.

So much for the language of borders/boundaries in STS. But how do these relate to the borders/boundaries of geography and politics, the disciplines most overtly concerned with borders and boundaries?

Borders in Political Geography

In his paper *'The Dynamics of Border Interaction: New Approaches to Border Analysis'*, historian Oscar Martinez lists some of the properties that have previously been proposed to differentiate border models. Martinez proposes four models of borderland interaction, taking into account environmental and social conditions that promote or inhibit cross-boundary ties. The models' description below is adopted from Martinez and is visualized in Figure 8.1. In the following Figure 8.2 these models are applied to the STS models as discussed above.

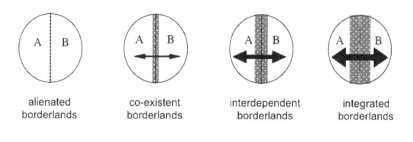

A

Figure 8.1: Four border models; adapted from O. Martinez, 'The Dynamics of Border Interaction: New Approaches to Border Analysis', in C. H. Shofield (ed.), *Global Boundaries* (London: Routledge, 1993), pp. 1–15. These A-models visualize the transgression potential of the border types in terms of intensity of relationships and material exchange modalities.

A-Model 1: Alienated Borderlands

This model of a border is one that is closed and sharply drawn. Boundary exchange is practically non-existent owing to extremely unfavourable conditions such as intense nationalism, ideological animosity, cultural dissimilarity and so on. Trade and person-to-person contact are difficult, if not impossible, and there is a tension-filled climate that keeps these areas unstable, sparsely populated and underdeveloped.

A-Model 2: Co-Existent Borderlands

Co-existence arises when there is minimal border stability, that is, when international relations are possible, for example, and economic and social life can attain a certain level of normality. For example, the two parties might reach a general agreement regarding the location of their common border, but leave unresolved questions of ownership of valuable natural resources in strategic border locales.

A-Model 3: Interdependent Borderlands

A condition of borderland interdependence exists when a border region is symbiotically linked with the adjoining region of another country. International relations are relatively stable here and there is a favourable economic climate that enables the whole borderland to enjoy growth. Eventually sufficient mutual commitment develops to allow the partners to become structurally bonded to each other. Economic interdependence creates many opportunities for establishing social relationships across the boundary, 'allowing for transculturation to take place'.[26] Interdependence implies that (at least) two more or less equal partners willingly agree to contribute towards and extract from their relationship in approximately equal measure.

A-Model 4: Integrated Borderlands

This is a stage where the partners eliminate their political differences and remove the barriers to trade and human movement across the mutual boundaries. There is a merging of economies and a diffusion of technology, and products and labour flow from one side to the other. Each partner willingly relinquishes its sovereignty to a significant degree for the sake of achieving mutual progress. Ideally the level of development is similar in both societies, and the resulting relationship is a relatively equal one.

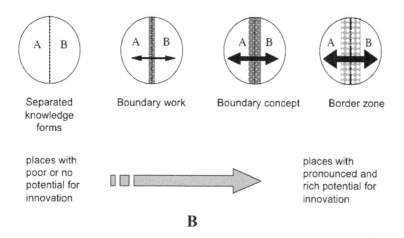

B

Figure 8.2: Knowledge building on the border; adapted from Martinez, 'The Dynamics of Border Interaction: New Approaches to Border Analysis'. These B-models apply the geographical-political models to STS concepts.

B-Model 1: Separated Knowledge Forms

This first model was described as having the most closed and sharply defined borderline. With practically no exchange across that line, opponents were strictly separated by tensions and ideology. The boundary work described by Gieryn – the separating out of different practices that then resulted in physics and engineering – brought about this kind of alienated borderland. It would probably not be too difficult to find more cases of this kind, especially in classical modern science through the nineteenth century – the boundary between primary and secondary qualities, for example, or between primary and secondary causes. The more interesting question would be whether the current scientific landscape still produces such borders or whether the extension of experimentalism undermines such divisions.

B-Model 2: Boundary Work – Co-existence

This model best represents the process of boundary work – though not its result. Creating a situation of normality is an important issue here. It is perfectly expressed in the activity of building images and methods that help to stabilize the border. Scientists from different disciplines working on different scales and problems can eventually agree on a computer model that represents both of their 'realities'. Now 'the two parties might reach a general agreement regarding the location of their common border'.

B-Model 3: Boundary Object – Interdependence

In this model, interdependencies of various kinds create many opportunities to establish social and economic relationships across the boundary, 'allowing for transculturation to take place'.[27] This corresponds to the requirements of the construction of the boundary object, with its ideal of cooperation among actors who remain rooted in their respective disciplines. Here, too, people come together for a limited period of time and are willing to act as more or less equal partners. They occupy a stable social world during this time, and eventually extract from it research data or objects.

B-Model 4: Border Zone – Integrated Borderlands

Kohler's border zone corresponds to the properties of B-model 4. Mixed practices and ambiguous identities coalesce with the model's merging of economies, diffusion of technology and free flow of products and labour on the border. Unlike the relative equality of the relationship between the two integrated borderlands, the relationships and tensions between field and laboratory science are never resolved, and ecology always remains a 'patched' discipline.

In conclusion, it can be acknowledged – hardly surprisingly – that borders/ boundaries are a fundamental condition for the constitution of objects: without borders and boundaries there can be no entities. Further, concepts of borders/ boundaries are necessarily ambiguous concepts: they form a link between two semantic fields but also create tensions. The surge of border/boundary concepts in the late 1980s reflecting the dynamics of unruly scientific research practices can be construed as a critical response to the definition of absolute borders and limits to development set in the 1970s. The hybridization of borders/boundaries did not just further expand the space for experimentation but simultaneously explored the blurring and transgression of seemingly fixed boundaries and thus the possibility for alternative scenarios that defy the limits to development. The primarily negative and limiting notion of border/boundary was gradually transformed into a positive notion of opportunity, with limits framed epistemologically relative to an ever expanding state of technology. The morphology of borders allows us to trace this move to an ever more clearly defined conception of the domain of ecology and thus of experimentation and knowledge construction in ecology and beyond.

9 EXCURSUS: 'BRIDGING SCIENCE' OR 'PROBLEM-BASED SCIENCE'?

In the following a case study is presented that elaborates 'aquatic ecology', analysing its transgressive character as a domain of research located somewhere between the social and the natural sciences and between the more applied and more theoretical domain. It seems that aquatic ecology is caught up in a permanent oscillation of basic concepts (or paradigms) involving experimentation with concepts and practices.

One of the co-founders of the International Association of Theoretical and Applied Limnology, August Thienemann, was quick to recognize the interrelationship between politics and science and between industry and the state in the sphere of ecological research. He sought to make use of it not only in conceptual terms but also as a tool for institution building. Thienemann wanted limnology to be seen as a 'bridging science' (*Brückenwissenschaft*) and to be recognized for 'its great cultural significance for our times'.[1] Thienemann borrows the notion of a 'bridging science' from a text that was programmatic not just for limnology as an ecological science. *The Structure of Wholes* (*Die Struktur der Ganzheiten*) was written by philosopher Wilhelm Burkamp (1929) and is repeatedly referenced by Thienemann. What seems capture Thienemann's imagination in particular is Burkamp's characterization of the new sciences as problem-based, methodological and factually structural wholes. This supports his own conception according to which the unique character of limnology is grounded in 'the [research] object and in the methods'.[2] According to Burkamp, 'the isolation or closure of sciences is a danger that can be mitigated by the numerous interspersed marginal and bridging sciences'.[3] It is in the following paragraph that Burkamp elaborates on the new sciences as problem-based activities that form structural entities with 'particular methods and perspectives'.[4] Thienemann refers to precisely this conception by proposing that a science like limnology could be one of those new interspersed 'partial sciences': 'The young science grows out of the ambition to address explicitly the external relations that became stunted in the old sciences'.[5]

By stating this view of the problem-based sciences,[6] Thienemann was effectively placing limnology in a mediating position between various scientific disciplines, predominantly hydrography, biology, geology and oceanography, as well as between basic and applied research. 'Theory will always remain the foundation for practice!' he exclaims, adding as a caveat that nonetheless

> the study of wastewater (an extreme milieu with regard to its chemistry), which was originally undertaken exclusively for practical reasons, has made it quite considerably easier for us to gain a theoretical understanding of the population in chemically normal natural bodies of water.[7]

What is especially remarkable about this statement is that, rather than supporting the familiar idea of knowledge transfer from theory to practice, or from the natural sciences to the engineering sciences, it also recognizes the reverse trajectory and acknowledges it as an important heuristic. Thienemann thus offers not only methodological arguments for the bridging function of limnology rather than justifying it solely with regard to a specific set of research objects; he also argues that the conceptual framework of this new science enables it to function as a bridge. He brings into play the link to concepts of natural philosophy, taking up a romantic conception of nature; at the same time, though, he also mentions modernity along with the progressive and experimental character of limnology, which he claims makes it different from the 'merely' descriptive approach of natural history.

Thienemann's own research also testifies to this breadth of scope: it was situated initially in both the theoretical and the applied sphere. And although his later research can be described as rather more theoretical in the conceptual sense, he remained active in fisheries biology, in relevant committees and organizations, and reference was frequently made to his biological studies of wastewater.[8]

In his later work, Thienemann uses the term 'bridging science' to characterize not only limnology but ecology as a whole, whose purpose, as the study of nature's household, is to connect 'all the branches of the study of nature'.[9] In the obituary he wrote for Thienemann, G. H. Schwabe stresses that the scientist's legacy lies in seeing the core task of ecology as one of nurturing connections – of acting as a bridging science not only between scientific disciplines but also between the natural sciences and humanities and thus between small and large interdisciplinarity, so to speak. Following Thienemann's example in this respect, Schwabe situates himself in the tradition of a holistic world view and sets his sights on founding an ecology that makes meaning and that he considers capable of preventing 'modern civilisation from drifting without an anchor into the incomparable'. He continues: '[E]cology is a logically necessary connecting link between the natural sciences and the humanities and as such is inescapably at the mercy of the tensions and conflicts that are in keeping with its very essence'.[10]

As if Thienemann's example (with support from Burkamp) was not powerful enough to illustrate the peculiar character of ecology, another metaphor of his extends far into present-day discussions of ecology and the field ideal of experimentation. This is the metaphor *Grenzland Limnologie* (borderland lim-nology)[11] which is meant to express the idea that limnology inhabits a space between the 'motherlands' of biology and physiography. Much like the term 'bridging science', the notion of a 'borderland', or border zone, or even 'border science' (*Grenzwissenschaft*) has become part of the terminology that serves to describe ecology today. As discussed above, historian Robert Kohler recently offered the concept of 'border zone ecology', in which objects, concepts and individuals constantly travel back and forth between lab and field and which thereby serves as an exemplar of contemporary research.[12]

The Work of Ecological 'Holism' in a Malleable Border Zone

In the preceding paragraph the different kinds of epistemic and ontic tracks were discussed that may appear in a scientific field operating in a border zone. As we have seen, Thienemann refers to limnology in his descriptions of ecology. This is also the case when he proposes the conceptual scheme of the three stages of ecol-ogy, which he presented in his 1942 paper 'On the Nature of Ecology'. This, too, is an attempt to give ecology a conceptual framework, being at that time a field still relatively untrodden in scientific practice, and using limnology as a model of an integrating science; this he could do because limnology was already institution-alized and possessed recognizable contours in the form of research programmes.

Thienemann was careful to anchor the programmatic element of his 'limnol-ogy' within a bourgeois educational canon, which included references to natural philosophy inspired by Goethe and a set of Christian motifs.[13] This involves an encounter between cosmological and romantic – even sacred – elements. There is much mention, for example, of the 'contemplation of the small with a view to the large', assertions that 'the whole always exists prior to the parts' and that 'the parts are a world unto themselves, and yet intermesh with one another harmoniously'. In the meantime the lake is described as an 'arena of life' or as a 'world in miniature'. Thienemann repeatedly emphasizes the connection to Goethe – thereby affirming a familiar topos and historical continuity that defies all political and epistemic breaks. In 1939, for example, the 'primary axiom of the holistic world view' is 'verified' by means of a quote from Goethe: 'How everything weaves itself into the whole, one in the other works and lives'.[14] The same sentence by Goethe is also quoted in 1951, again with reference to the 'axiom of holism'; indeed, the 'intellectual world of Goe-the also lives on in today's natural science, according to the authoritative judgment of modern scientific researchers'.[15] As a final example, in 1954 we find this:

> Our intellectual attitude towards nature must not be exhausted in the endeavour to identify its laws as the foundation of our material culture; instead, more than this, it must strive upwards to encompass a view of nature such as the one into which Goethe breathed life, who saw in living nature and in every one of its parts the whole, the

great harmony in all disharmonious separate phenomena, and to whom it was a well-ordered whole, a cosmos.[16]

This view was matched by an ideology of the nation state. Thieneman writes, for example, that 'the biological foundation for shaping and consolidating our German world view must be the science, which explores the great interrelations in living nature, namely: general ecology'.[17] Almost twenty years earlier limnologist Erich Wasmund (1902–45) had written that 'language is a part of the style of the innermost being of a people, and thus its specific scientific type becomes essential to it. People and soil create a material conditionality of their scientific types'.[18] Here a further aspect of holistic conceptions becomes clear, namely, that they serve to elide the political dimension of the concept of nature. Nature is treated as a moral category and is invested as such with an authority that cannot be betrayed. Above all, though, whole and harmonious nature becomes an icon of a way of thinking rooted in cultural pessimism which, during the Weimar period, became an integral part of bourgeois self-perception and that was further fuelled by, for example, Oswald Spengler's book *The Decline of the West* and by the philosophical anthropology of Max Scheler, particularly his piece entitled *The Place of Man in the Cosmos* (*Die Stellung des Menschen im Kosmos*), first published in 1928.

According to this cultural pessimism, humans are perceived above all as destroyers of nature; at best they assume the role of benevolent stewards, but nature can definitely not be a discursive political term. This position – critical of modern civilization and often technophobic as a result – was widely held in early ecology as well, where a more or less explicit commitment to a holistic image of nature played a central, integrating role and made up the philosophical core of the discipline. Yet such a sceptical position is by no means necessary even to a science that is committed to holism, whether conceptually or politically.[19] Accordingly, the understanding of holism in the field of ecological knowledge plays out in very different ways: it may refer primarily to practical knowledge, as envisaged by the planning sciences and scientific nature conservation; or it may be elaborated in a more systems theoretical way, as with the ecosystem models of 1950s American new ecology; or again it may present itself as extremely technophile as with the large-scale experiment Biosphere 2.

10 ECOTECHNOLOGY COMPLEMENTS ECOSCIENCE: PROBING A FRAMEWORK

In many societies a growing consensus has arisen about the importance of ecological knowledge. More and more people are coming to understand that such knowledge can help to address some of the most pressing problems we face at both global and regional levels. It can do so not only by providing research and effective management options to deal with global warming, the depletion of natural resources and the deterioration of soils and water resources, but also by underpinning widespread calls for greener systems of production and consumption, in industrialized as well as newly industrializing countries. Debates about natural disasters, about the purity of nature and about the perceived crisis of the nature-culture relationship in general have one thing in common: they revolve around the important role of ecological knowledge in social processes and in negotiations over the kind of environment in which we wish to live. Accordingly, institutions engaged in environmental and ecological research are believed to have the necessary expertise to tackle environmental concerns and to develop solutions. In many countries, environmental research has become firmly established, whether in the context of implementing conservation programmes for water and land resources, establishing national parks and ecosystem services, designing environmentally friendly products, or promoting emerging technologies and green lifestyles. 'Environmental research must turn its attention to developments in the field of innovation in order to be able to contribute towards precautionary policies'[1] has become a widespread credo. Nowadays, environmental research is one of the innovative forces in modern societies and is no longer identified with the rather technophobic ideas and practices of the 1960s and 1970s when environmental movements were emerging and gathering strength in the US and Europe.

Thus, as part of an evolving narrative, 'eco' shifted from being a proxy term connoting nature conservation and scarcity alongside theories of balance and cycles (all tinged with a rather technophobic undercurrent) to becoming the herald of an enthusiastic culture of innovation and of practices of abundance[2] and squandering along with a moral economy of hedonism.

To begin we shall explore what kind of experimental practices are gaining significance in the broad field of ecological knowledge production and how these practices intervene in the rich interplay of 'representation and intervention', particularly regarding research technologies, instrumentation, experimental techniques and modelling practices. This will be discussed according to a conceptual framework that views the vast field of ecological knowledge through the lens of a juxtaposition of two umbrella terms, ecotechnology and ecoscience.

Ecotechnology will be discussed mainly in relation to the world of sustainable development and its practices and conceptions, which are explored from several perspectives using a range of conceptual tools. In this general sense, ecotechnology is understood as a technoscience that principally develops local theories and practices. It can be regarded as an instance of applied research that encompasses the engineering sciences – or, in the terminology preferred in this book, as 'use-inspired basic research' (following Stokes's four-field scheme[3]). Good examples of this include restoration ecology, ecological engineering and industrial ecology. It is interesting to note that explicit reference has been made to the latter in the context of recent activities around so-called green nanotechnology. Industrial ecology is believed to afford a framework for the sustainable development of nanotechnology.[4] This will be discussed later in more detail; for the moment, it is the complementarity of ecotechnology and ecoscience that will be explored. Drawing again on Stokes's system, ecoscience can be regarded as an instance of 'pure basic research', that is, the search for basic understanding with no interest in application. However, there is another field, the 'double no' field, which might also apply here; it is a field concerned with neither basic understanding nor application. With its almost cryptic question mark, this field could be filled, for instance, with the collecting and reorganizing of data as practised in the relatively new domain of data mining or in the material sciences. Or again it might contain the activities of biologists conducting research in biodiversity or (further back historically) natural history. Although such an allocation assumes rather a lot (such as that the collecting activities done by natural historians were not driven by colonial interests regarding resource exploitation and therefore by an interest in application), it illustrates that there is a type of scientific activity that seeks understanding and yet is not identical with the search for basic laws or causal relationships. Instead, such an activity seeks basic understanding through the principle of resemblance, namely, by drawing on iconic relations. Philosopher William Whewell spelled out this principle in the methods of gradation and natural classification, defending these methods vigorously as being on a par with the application of the mathematical sciences.[5] In accordance with these methods, ecoscience is characterized by the development of general concepts and theories, something that is done in theoretical ecology, for instance, which has generated the competitive exclusion principle, models depicting predator-prey relationships and ecosystem theories.[6] Ecoscience also includes systematic work on biotopes and plant/animal communities, on ecophysiology, and on parts of hydrology and geology, e.g. studies of ion exchange in soil or the dynamics of turbulences in running water. Ecotoxicology issues may also be addressed from an ecoscientific perspective rather than from an ecotechnological one. And if we understand ecoscience in a more general sense as the science of organisms and their environment and of perception and action (that is, of modes of connectedness), then all kinds of hybrid domains between ecology as a natural science and the social sciences and humanities begin to emerge, such as ecolinguistics or ecological psychology. In sum, one might say that ecoscience seeks to overcome the dimension of singularity and instead to describe rules of connectedness using more general concepts, models and sometimes even laws.[7] This can be seen in distinct contrast to ecotechnology, which is about developing tailored solutions and site-specific practices.

Before discussing the experimental practices in ecoscience and ecotechnology and how they help to intervene in the world of environmental concerns, an initial snapshot will be taken of the manner in which the field of ecological knowledge emerged. Having presented a complementary conception of ecotechnology and ecoscience, it seems almost superfluous to point out that ecological knowledge is not generated in a single scientific discipline called 'ecology' but that it is more adequate to conceive of ecology as an 'interdiscipline', that is, as a form of knowledge production that is scattered over several academic disciplines. What this requires is, first, precise philosophical reflection regarding the prevailing plurality of methods, objects and concepts in this field and, second, focused epistemological effort to describe this patchwork field. This will be done in the following by discussing scientific ecology as an ecoscience, always keeping in mind its complement ecotechnology.

Ecosciences: Disciplining Interconnectedness

In the following this ecoscience 'ecology' will be discussed in terms of presenting a 'snapshot' from the history of human-environment relationships. Scientific ecology is different from the environmental sciences in that it is the scientific mode of depicting the relationship between humans and the environment. In other words, ecology is just one stage in the history of scientifically conceived human-environment relationships.

The term 'ecology' was introduced in terms of being the science of the 'household' of organisms, that is, of organisms' relationship to their biotic and abiotic surroundings. This naming occurred in 1866, when 'Oekologie' was introduced by German zoologist Ernst Haeckel, who chose it to refer explicitly – like 'economy' – to the Greek 'oikos', or 'household'.[8] However, the new name 'Oekologie' paved the way for a new self-conscious enterprise in a number of nation states and linguistic regions simultaneously. The forerunners of ecology were certain branches of natural history concerned with geopolitical and resource-oriented knowledge in the colonial and imperial era of the seventeenth and eighteenth centuries. Around the 1850s these were followed by a scientifically oriented geography that claimed to provide the most comprehensive account of the human-environment relationship. This involved a particular form of the idiographic method, namely, a scientific perspective on the individuality of a form or place, conceived of as a manifestation of the universal in the particular, the particular being the measure of validity.[9]

It was not until the post-WWI period and, much more prominently, the post-colonial Cold War period in the twentieth century that ecology fully took hold and ecological knowledge became an important factor in the environmental crisis and the reorganization of modern societies.[10] Now, in the twenty-first century, ecology appears to be largely replaced by a steadily evolving domain of environmental sciences. However, it is highly likely that the new field of environmental sciences will not continue to represent the primarily academically oriented perspectives of geography and ecology. Instead they are more akin to practically oriented fields such as the agricultural sciences, forestry, fisheries, etc. and indeed go beyond these in a manner that might be called technoscientific or post-normal[11] – indeed ecotechnological. Another way of characterizing the border crossing character of environmental issues was – and still is – in terms of transdisciplinary research (noting especially the interaction between academic and economic and social concerns and thus constituting specific conditions of knowledge production). Meanwhile, the term transdisciplinarity has become firmly established, a fact that is reflected, among other things, in its inclusion in the *Encyclopaedia of Life Support Systems* (*EOLSS*), where it is described as follows:

> Socio-ecological transformations at global, regional, and local level are defined as the general objects of this new type of research. The scientific and technological knowledge needed for an understanding of these transformations is distributed over a broad spectrum of disciplines and professions committed to incommensurable values, different theoretical concepts, and conflicting methodological orientations. Therefore, a strong demand for integrated knowledge has arisen with the aim of improving both its explanatory power and usefulness for problem solving.[12]

As mentioned at the start of the present chapter, this 'new mode' of ecotechnologically managed human-nature relationships tends to be local and action-focused and permeates different domains of society. It produces knowledge that is based on consensual and democratic demands rather than on scientific standards of universal or timeless validity.

'Ecology' as a Snapshot in the History of Human-Environment Relationships

In this book, it has been claimed repeatedly that the construction of ecological knowledge is to be described more in terms of a field of knowledge than as a single discipline and that it is pluralistic in many ways – theoretically, institutionally, in terms of its practices and with regard to its background assumptions, with their more or less obviously ideological, political, sceptical or metaphysical character.[13] In the following it is argued that the existence of a plurality of theories and programmes of experimentation in ecology as an ecoscience has a positive impact insofar as it facilitates greater logical flexibility and thus also greater explanatory power.[14] The question to be answered is how this scientific ecology can be characterized by reference to forms of generalization and how these forms are supported by experimental practices.

From the point of view of a general philosophy of science, this prompts the question of how such a framework – one that is open to pluralism and is not constrained by a strong commitment to unified science or even monism – is to be conceptualized. An increasing number of proposals relating to scientific pluralism have been put forward over the last twenty years. Among their more well-known exponents are Nancy Cartwright, Ian Hacking, Helen Longino and Alan Richardson, while their forerunners are often said to include Patrick Suppes, Alfred North Whitehead and William Whewell; even Karl Popper and Paul Feyerabend agreed on the point that '"(t)heoretical pluralism" is better than "theoretical monism".'[15] According to Helen Longino, then, it is not only admissible to produce partial knowledge, it is even acknowledged and accepted that these partial knowledges do not operate as parts of one explanatory machine, as in a neat division of labour where every question is first approached with its own distinctive set of methods and finally cobbled together in a synthetic manner. Instead, there are rather different knowledge claims and modes of explanation at work (conceptual, causal, model-driven, data-driven, etc.). Longino argues that 'the multiplicity of approaches is usefully addressed not by comparative evaluations directed at selecting the uniquely correct one, but by appreciating the partiality of each ... In concert, they [the approaches] constitute a nonunifiable plurality of partial knowledges.'[16]

The present study aligns itself with Longino's comments and with similar arguments for tolerance towards different knowledge claims and multiple relations within and among disciplines. It also adopts the approach proposed by Rudolf Carnap, who accepts that the disciplines have their own intellectual life, given that they are subject to philosophical considerations:

> Let us grant to those who work in any special field of investigation the freedom to use any form of expression which seems useful to them; the work in the field will sooner

or later lead to the elimination of those forms which have no useful function. Let
us be cautious in making assertions and critical in examining them, but tolerant in
permitting linguistic forms.[17]

Focusing on the narratives generated by the philosophy of ecology (and, in a
broader sense, of biology), there are a number of approaches in circulation which
offer some ideas about defining generality. All of them avoid the classical deductive scheme, placing an emphasis instead on inductive operations and the role of
scientific practices in ecological research on connectedness. These studies share a
certain degree of reservation towards the debate on scientific laws as a universal
principle for all sciences, calling instead for a more sophisticated debate on rules
and generalizable features – very much in keeping with Carnap's appeal to permit different linguistic forms.

In a first snapshot we shall look at Gregory Cooper's threefold scheme in
which he proposes theoretical principles, phenomenological patterns and causal
generalizations as the basic forms of generalization in ecology. This constitutes
a philosophical taxonomy which is intended to functioning as an aid to distinguishing the different modes of investigation and acknowledging their varied
generalizations while not dismissing the possibility that laws may exist in ecology.[18] Cooper's concern is to find a conceptual tool that recognizes degrees of
contingency in ecological generalizations; this is in order to cope with the fact that
contingency does not always play a role in ecology. Ultimately he suggests that 'the
attempt to partition generalizations into the two categories of laws and nonlaws
should be abandoned in favor of the concept of nomic force ... which recognizes
that nomicity in biology comes in degrees and over restricted domains'.[19]

Kenneth Waters is also wary of the concept of law. He explores a conceptualization of biological generalizations in which a distinction is drawn between two
types of the latter: the first refers to the distribution of traits among biological
entities such as populations or groups, while the second describes dispositions
of causal regularities.[20] The trade-off in this distinction is that the evolutionary
contingency hypothesis only applies to some biological generalizations, mainly
the first type (that of distribution). However, generalizations about biological
regularities can be extended to distinct systemic classes which are independent
in time and space.

Similarly, Sandra Mitchell argues that the dichotomous oppositions 'law vs.
accident' and 'necessity vs. contingency' are, if anything, an obstacle that impov-
erishes the conceptual framework of the biological sciences. Accordingly, she
defends a 'multidimensional account' of scientific knowledge, linking in with
the plurality debate and proposing a multi-dimensional conceptual space that
spans the axes of abstraction, stability and strength. This scheme allows for a
more comprehensive conception of 'law' in which both the law of conservation

of mass and Mendel's law can be represented in the same conceptual space, with the latter simply being less stable. The advantage of this clever model is that 'the strength of the determination can also vary from low probability relations to full-fledged determinism, from unique to multiple outcomes'.[21]

A final snapshot takes us back to the philosophy of ecology and to the role of practices in ecology. Kristine Shrader-Frechette and Earl McCoy argue that the main method used to link data with a hypothesis (in case studies, for instance) constitutes a variant of logic which they call 'informal inferences'.[22] These informal inferences also include pragmatic procedures, which ultimately makes them a reliable basis for comparison as well as generalization, as the authors themselves point out. Shrader-Frechette and McCoy hold that the rationality of practice relies on rules and on the reference to a community of individuals, just as scientific concepts themselves do.

> Hence the 'logic' of case studies may be appropriate to science if one conceives of scientific justification and objectivity in terms of method, in terms of practices that are unbiased – rather than in terms merely of a set of inferences, propositions that are impersonal.[23]

Thus the uniqueness of case studies must be seen to lie not only in their enactment of purely subjective rules and unverifiable principles. Instead, ecological case studies should be perceived and appreciated epistemologically for the specific form of knowledge they manifest, which accommodates practical as well as conceptual and methodological analysis. To put it in a nutshell: it is here where 'experiments in practice' become fruitful for ecoscience.

The approaches discussed above – be they Cooper's concept of nomic force, Mitchell's multidimensional space or Shrader-Frechette and McCoy's proposal of a 'logic' of case studies based on pragmatic procedures – all provide appropriate instruments for describing ecology as a powerful scientific discipline possessing a capacity for generalization. Furthermore, our discussion – particularly of the approach taken by Shrader-Frechette and McCoy – shows not only that the analysis of ecological practices plays an important role in better understanding the epistemological features of case studies but also that ecological practices may themselves contribute to the formulation of general rules in ecology. Gaining a better understanding of the practice of experimentation in ecology, be it 'nature's experiments', 'mesocosm studies', 'quasi- experiments', 'real-world' or 'social' experiments, thus also helps to better understand ecoscientific patterns and to strengthen the complementarity of ecoscience and ecotechnology and their different modes of knowing.

Object Hedonism and a Plenitude of Practices and Materials

In the foregoing section it was pointed out that the positive appreciation of a plurality of methods and theories is not only indispensable in ecology but enhances its explanatory power.[24] This is reflected in its choice of objects, which appears to be rather pleasure-driven: ecology deals with macroscopic and microscopic plants, with all kinds of animals, with microorganisms and with systematically indeterminate objects such as fungi and lichen. These species or systematic groups might populate mountains, lowlands, lakes or oceans, but they may also be found in old tree trunks or in caves.[25] Obviously, ecology does not partake in the usual division of labour that is prevalent in the biological sciences and that in most cases draws on a combination of systematics and relative dimensionality: botanists, zoologists and physiologists deal with macroplants and -animals, plankton and pollen specialists with microplants and -animals, microbiologists with bacteria and protozoans and, finally, molecular biologists with macro-molecules.[26] This division of labour is tied to experimental methods of isolation and purification, to instruments and to techniques of preparation and representation: scientific objects need their own environments and modes of treatment to become visible, tangible and ultimately detectable in the scientific context. This applies not only to the micro-scale, where biologists necessarily have to engage with instruments such as microscopes, microtomes or mass spectrometers in order to 'realize' their objects, as Gaston Bachelard pointedly formulated.[27] It is also true for ecological macro-scale objects such as trees, reptiles or insects. All these objects must first be located – 'seen' from a scientific point of view – and then prepared and preserved in a methodologically prescribed manner; they need to be controlled and normalized.

The experimental practices involved in ecological research, whether in the field or in the lab, have rarely been investigated from a science studies perspective. Bruno Latour, who is known predominantly for his laboratory studies, presented a study on a changing ecotone in the tropical rainforest in Costa Rica,[28] describing in some detail the practices and material transformations involved in this research while advocating his 'chain of inscriptions'. Here, the material point of origin was a pile of soil, taken from a particular vegetation border zone in Costa Rica's jungle, and stretches – with no break whatsoever in the chain of material transformation – to a printed and 'tradable' representation of this place, having been transformed into an ecotone.

In a similar vein, Wolff-Michael Roth conducted a series of detailed analyses on research practices in ecological field experimentation involving, among others, lizards and salmon.[29] In the field of biodiversity research, Ayelet Shavit and James Griesemer consider the issue of locality and the problem of replication. Using remote sensing techniques when revisiting a species' locality, they argue, does not offer a 'general, rational solution'.[30] Instead they advocate a scale-dependent 'resolution' along with a crucial role for the idiographic method.

Finally, in this admittedly incomplete list of references, the Field Book Project[31] is almost a must with regard to the role and plenitude of inscription devices of field sciences. The project is a joint initiative of the Smithsonian Institution Archives and the National Museum of Natural History in Washington, DC. Established in 2010, its aim is to create a field book registry, linking collection and archival work with research. Field books are inscription devices, as are their better known cousins, laboratory journals, and they come in manifold forms: in addition to journals, specimen lists and species accounts there are also scrapbooks, photo albums, logbooks, correspondence and other primary source materials. In his fascinating collection entitled *Field Notes on Science and Nature*,[32] Michael Canfield offers some interesting thoughts on record keeping in the field. The technologies used for this might be just paper and pencil or they might be digital devices and linked-up databases. In either case, decisions need to be made when thinking about what the form affords and – in the case of digital versus analogue – 'what may be gained or lost with the implementation of digital notes'.[33] The range of field sciences presented makes the book a rich and 'thick' collection, ranging as it does from natural history to the modern sciences of geology and ecology, and from social anthropology to archaeology. A variety of heuristic uses and memory techniques are proposed and discussed, partly with respect to ecological case studies. One contributor to the collection points out that the field journal method allowed him to browse through his memories and 'to select which incidents can provide a core around which a publishable account can be constructed';[34] in other words, journal inscriptions not only helped in putting forward a hypothesis but also contributed towards theory building. A potentially worthwhile future project might be to pay more attention to this process of theory building as it emerges from a 'frozen' state in the narrative records of memory techniques and modes of practice. A closer look at these recording tools and how they connect to scientific practice and theory building may certainly help to shed more light on the elements identified that characterize experimentation in the field.[35] They undoubtedly provide insight into the material and conceptual transformations of the emerging object.

In conclusion, what all these material studies have in common is that they demonstrate strikingly just how much work and negotiation is required to make the scientific field experiment happen and thus to bring to realization the ecological object in the field and to make it a matter of concern.

Ecotechnology: On Connections in Space

Usually it is industrial ecology and bioengineering along with restoration ecology and the sustainability sciences that are considered to be ecotechnologies. However, following a historically and epistemically more informed scheme, cabin ecology, space ecology and even green nanotechnology can also be aligned with these domains, insofar as they are concerned with space. It is important to note that this space should be understood not simply as a geometrical dimension but rather as a dimension that refers to technological capabilities and to an enabling capacity. Accordingly, it is a cognitively and instrumentally bounded space full of promises and potentiality.

This concept of space has been defended mainly in relation to the technosciences, to which, no doubt, all of the above mentioned domains can be assigned. The question arises, then, what makes them especially 'eco'-technoscientific? The eco-argument is based mainly on space as an oikos, that is, as a surrounding space that affords the organism *Homo sapiens* its habitable oikos. This sounds rather banal given that it is, of course, also the programme of technoscience to improve the conditions of human life through innovation: *Homo faber* has become a fairly common species. It is this permanent process of reform of knowing and manufacturing that Hannah Arendt refers to when she places such great emphasis on 'fabricating experiments' (*fabrizierende Experimente*), as she calls them; at issue, for her, is the making of an artefact, of a 'work' and, more generally, a shift from asking 'what' and 'why' towards asking 'how'.[36] Arendt points out that it is the success of technology and science and, particularly, of their alliance – thus of technoscience – that bears witness to the fact that the act of producing or manufacturing is inherent in the experiment: it makes available the phenomena one wishes to observe.

However, it is rather Arendt's *Homo laborans* who inhabits the ecotechnological world, a world in which an exuberance of energy and materials and the relentless production and consumption of goods is the driving force.[37] All these largely industrially produced artefacts (cars, domestic appliances, hardware, etc.) must be consumed and used up as quickly as possible lest they go to waste, just as natural things decay unused unless they are integrated into the endless cycle of the human metabolic exchange with nature. 'It is as though we have torn down the protective walls by which, throughout all the ages past, the world – the edifice made by human hand – has shielded us against nature'.[38] Arendt offers here a vision of the human-environment relationship that copes perfectly with the rather pessimistic world view of environmentalists in the late 1960s. She sounds an ecotechnological warning when she cautions that the 'specifically human homeland' (*spezifisch menschliche Heimat*) is endangered mainly because we erroneously think we have mastered nature by virtue of sheer human force, which is not only part of nature but 'perhaps the most powerful natural force'.[39]

The key feature of ecotechnologies lies in the relational character of the connectedness they manifest: on processes rather than dispositions, on the oikos instead of physical space. Technological and social innovation is needed because the relationship between humans and their environment (mainly the material environment, whether artificial or not) is not yet sufficiently developed and needs to be optimized – complying with whatever social and political values and norms exist in the locale concerned. In his grammatical exposition *On Settling* (as opposed to striving), philosopher Robert Goodin has termed this process of balancing the relations between humans and their environment a process of 'settling in' which, in part, 'is a matter of adjusting yourself to your new place. In part it is a matter of adjusting the place to you'.[40] In a similar vein, historian Thomas Hughes points out that 'we' (humanity) have failed to take our responsibility 'for creating and maintaining aesthetically pleasing and ecologically sustainable environments' and that we should finally accept our responsibility to design a more 'ecotechnological environment, which consists of intersecting and overlapping natural and human-built environments'.[41] This appeal is addressed mainly to engineers, architects and environmental scientists whom Hughes considers to be the experts suited to design and construct 'the ecotechnological environment'.

It can be noted that space (and place) as an oikos in the mode of experimentation is the recurrent theme that links recent debates on climate change, green lifestyles, restoration ecology and industrial ecology as well as historically more distant issues such as the 'blue sky' campaigns (against air pollution in industrialized countries), efforts to combat water pollution in the nineteenth and well into the twentieth centuries, the management of the dying forests phenomenon and space ecology.[42] Accordingly, it is hardly surprising that the space-oikos theme developed mainly in the context of sustainability discourse, without always being explicitly spelled out. This work of conceptualizing ecotechnology through a better understanding of ecotechnological space is pursued in the following by moving diachronically as well as synchronically through different domains.

Restoration Ecology

Let us turn first to a concept – 'building with nature' – that is familiar in restoration ecology as well as in landscape architecture and ecological engineering, all of which are traditional domains of ecotechnological design. 'Building with nature' is used in more theoretical design approaches as well as in science policy and industry, where landscape services have become an economic good and 'building with nature' even a registered trademark, as indicated by a Dutch consultant's website 'Sustainable Development by Building with Nature®'.[43] 'Building with nature' is increasingly seen as a guiding principle and is usually accompanied by a rhetoric of sustainability and naturalness that is at once seductive and obscure – and not uncommonly infused with economic interests. In this respect it blends in readily with the broader societal movement towards 'greenness' which, in the case of landscape architecture, manifests in ever growing demand for urban green spaces, green corridors or even so-called 'new natures', that is, artificial pieces of nature that look like naturally developed landscapes located in the middle of cities or industrial zones.[44] Ways of reading the debate about 'building with nature' seem to be located between two extreme positions which can be characterized as 'conservatio' and 'mimesis'.

Each comes with a different historical and systematic background which ultimately manifests in different practices of building (with) nature. 'Conservatio' takes 'building with nature' to be a principle of preservation in which tradition and origin – whether of landscapes or other natural things – constitute an indispensable rationale. In contrast to this, the mimesis principle is understood as a principle of transgression, not simply in the sense of mere copying but as a process of creation out of which something new emerges. Building with nature here has something to do with the notion of 'using nature to go beyond nature'.

The first reading of building with nature, 'conservatio', is possibly the more traditional one. Building with nature according to the conservatio principle is teleological to the extent that the goal is predetermined by the reality of what is already present *qua* nature: its supreme precept is preservation/conservation, that is, caring for what has evolved historically, be it something that has grown naturally or as tradition. Accordingly, building is undertaken in order to protect and to preserve. This applies equally to the management of a nature reserve with its marvellous landscapes as it does to the inconspicuous forest pool that was once a bomb crater and is now a habitat for rare amphibians, or to pre-modern industrial heaths such as the *Lüneburger Heide* in Germany, with its traditional sheep farming. In each of these examples the overarching objective is to preserve as many elements of the historically and genetically produced landscape as possible, a landscape that need not necessarily be un-built up but may also be a cultural landscape. The aesthetic quality of a landscape is grounded in its genesis and in the relations it represents.[45] These historical landscapes are to be preserved in nature reserves and national parks; nature conservation areas are to be set up as museums with strict conditions of access – namely, for connoisseurs of such exhibitions. In the conservatio reading, then, 'building with nature' means encountering nature as an authoritative counterpart. That which has arisen in and out of nature has priority – *a priori* – over the technological works of human beings; the works of nature are morally superior.

The mimesis of building with nature is based on the idea that we might be able to go with nature beyond nature. This conception is not new, having a long pre-history in, among other spheres, alchemy and other magical arts as well as in more recent theories of self-organization and, not least, in bionics. In bio- and nanotechnology this idea is related extremely effectively to the molecular and atomic level. What is surpassed in this case is nature in the sense of the existing materiality of an organism or of a material by an abstract and law-governed nature such as that found in physics, molecular biology or carbon chemistry; it then becomes possible to envisage the prospect of developing organisms that are better adapted to industrial agriculture or materials whose manufacture involves achieving a more efficient economy at the atomic level. It is physically conceived nature that marks out the horizon of expectations foreseen by biotechnology or nanotechnology – an open-ended horizon, provided no contradiction emerges to the laws of nature. Just as open-ended, though, is the connection between the mimetic model of building with nature and the purported model of 'given nature': 'given nature' is not accepted as being, say, a regulatory idea or a restraining stipulation (as is the case with building with nature on the principle of conservation). The aim is instead to improve this same nature: materials – and even organisms – are no longer the original matter and accepted point of reference on which research is based. Similarly, structure-function relations are regarded as a hindrance: functions need to be liberated, as it were, from this set of relations. This constitutes a technical extension of nature – new materials which promise technological optimization and which open up new sets of possibilities. Thus in the age of technical reproducibility,[46] building 'with nature beyond nature' means enhancing and intensifying the nature in which we live, which we can touch with our own bodies and which is our environment in the ecological sense of the term.

The proposed mimesis reading of landscape also works with a figure of enhancement. Here too the idea is to outdo nature in its given form. In this case, though, it is not so much physical nature based on laws of regularity that provides the means to outdo nature when building with it, as described above for nanotechnology and biotechnology. Rather, it seems to be the principle of a technical and aesthetic construction of a concrete nature *qua* model that is manifested here. This new nature is constructed *through* the model of an ideal landscape and is built in this particular place and at this particular point in time – better than nature could ever have done so itself. The newly built piece of nature is an ideal type that unifies within itself an array of real places and times all at once and yet is a material reality, recognizable as a landscape. The gathering of conditions in particular places at particular times is ultimately what brings forth the ideal-typically built piece of nature. This reading of 'building with nature' is all about ecotechnological innovation: a new landscape is constructed using planning techniques and design elements; technological development and scientific research work hand in hand with one another.

Cabin Ecology

Cabin ecology is a particularly interesting historical case.[47] This serious scientific-technological research programme began in the 1950s as a field of research with legions of technicians and scientists working on the technical and conceptual implementation of water, nutrient and gas cycles. The story of the emergence of cabin ecology is closely related to the dream of developing outer space as an unlimited spatial resource by establishing human settlements in Earth's orbit, or even by colonizing Mars.[48] As early as the 1960s, a well-known ecologist noted that this kind of ecotechnological research was 'likely to draw more attention (and surely more money!) than biology'.[49] The technical conception of constructed ecosystems for space travel took on added significance when the entire planet became visible as a spaceship that needs to maintain conditions of life for a human population. The 1968 Apollo image of the blue planet brought into view not only the Earth as an enclosed and, above all, limited space but in addition the various scientific parameters for describing space (closed-loop cycles, stability, 'carrying capacity' and so on). 'Spaceship Earth' was no longer associated with space travel but increasingly with the emerging environmental discourse. The notion of limits to growth had become the central driving force, while the hitherto technical model of space for astronauts was turned into a macroeconomic model. The theory of carrying capacity and the limited nature of space as a resource were expanded to include other resources, a transformation advanced by economist K. E. Boulding. The technical model of space travel for astronauts was projected onto the planet as an object of management. In this way, Boulding turned the cabin or spaceship into a model in which carrying capacity played a major role and the limitation of space became identified with all other resource limitations:

> the earth has become a single spaceship, without unlimited reservoirs of anything, either for extraction or for pollution, and in which, therefore, man must find his place in a cyclical ecological system which is capable of continuous reproduction of material form even though it cannot escape having inputs of energy.[50]

Thus the 'spaceship' became the rational model for the global management of Earth, but one in which humans could suddenly turn into an irritant by producing too much CO2 or waste. Humans became a form of 'pollution' on Earth, spreading like a disease and putting Gaia in mortal danger, as ecologist James Lovelock put it.[51] This 'economy of Spaceship Earth' also underpins those models that provided scientific support for the concerns expressed in the Club of Rome report on the 'Limits to Growth', published in 1972. The economic models also laid the groundwork for the development of the first 'world model', intended to characterize Planet Earth using only a few parameters. The management of space as a limited and limiting resource had become a form of control of the spaceship and of the planet in general.

Although it had its heyday in the 1970s,[52] 'Spaceship Earth' continued to attract attention with spectacular projects such as Biosphere 2 in the Arizona desert, which is purported to be a simulation of Biosphere 1, 'our blue planet'. Originally the project was financed privately by multimillionaire Edward Bass and was backed up scientifically by ecologists Eugene and Howard Odum. What was meant to be a self-sufficient ecosystem failed in scientific and technical terms, because it was necessary to intervene in order to prevent serious harm to the eight 'bionauts' when the circular flow of elements did not function as expected. Although less well known, a third biosphere, based in Siberia,[53] exists in addition to Biospheres 1 and 2. It was built between 1965 and 1972 and is still in operation. A simulated journey to Mars can be seen as an even more recent addition to the cabin ecology programme.[54] Thus 'space' is not merely one of numerous physical parameters but provides an analytical baseline that is also a scientific object and technoscientific instrument of control in ecotechnology.

Green Nanotechnology

It appears that sustainability and nanotechnology discourse alike deal with space as a resource: both of them accept and incorporate arguments about limited growth and develop strategies of control in response that open up a boundless space – literally and metaphorically – of technical possibilities.[55] 'Green nanotechnology' portrays itself as a technology that is at once environmentally friendly and innovative. It turns nanotechnology's general promise of controlling matter and 'shaping the world atom by atom' into a green promise. It is noticeable that well before the term 'green nanotechnology' appeared on the public stage, the production of scenarios for sustainable nanotechnology was already under way – 'smokeless industries', 'wasteless products', 'limitless energy production' and 'full pollution control' are just some of the slogans that circulated in roadmaps and reports put out by government and industry stakeholders. The following quote gives a flavour of those visions:

> if the oceans could be used for growing biomass fuels or harvesting energy through nano-biotechnology advances, significant increases in global energy supplies would result. Systems life cycle thinking is particularly important in addressing the energy issue because of the coupling of energy and the environment ... properly designed, systems using such technologies as photovoltaics, engineered photosynthesis, factory process heat re-use, or agricultural fuel production could lead to a world of sustainable energy, agriculture, and climate.[56]

Proponents of green nanotechnology prefer to avoid stereotypical stories of dystopia and utopia, something which may already be part of the politics of the concept: sustainability aspects thus become absorbed entirely into green nanotechnology. The concept of 'green nanotechnology' must not become entangled with highly charged and severely criticized visions. Hence, the new label of green nanotechnology allows for a new positioning of arguments, still an ambitious

enterprise but one that is explicitly more 'real' in terms of products and market-
ing strategies and more feasible in terms of societal acceptance and governance
models. Even more importantly, this conceptualization involves a novel cognitive structure that ultimately moves nano discourse closer to certain elements of
sustainability discourse, namely, the embrace of both innovation *and* sustainability. Green nanotechnology thus facilitates a rearrangement of the conceptual
furnishings on the public stage. This can be readily observed using the example
of the US initiative on green nanotechnology, organized and supported mainly
by the Woodrow Wilson Center. The concept of green nanotechnology relies
explicitly on a proven facet of the sustainability concept being well established
already in green chemistry and engineering science. A certain emphasis is placed
on downstream measures aimed at regulation and control, which is also reflected
in the strategic package devised by the Woodrow Wilson Center.[57] This package
contains thirteen points that are formulated as direct recommendations, including: strict quality control over the identity of the nano products and methods
claimed to be 'green'; increased support for university research in the area of
'green nano' and collaboration with industry; use of federal capital for supporting 'green nano' products. All of this serves to portray green nanotechnology
as potentially fulfilling a 'twofold dream':[58] products can be constructed from
scratch in accordance with sustainability principles, and older products that are
at least potentially harmful to the environment can be replaced by 'greener' ones.

The significance of spatial conceptualizations is quite evident here. What
is also striking is the superimposition of the discourse of nanotechnology onto
the discourse of sustainable development. Both refer to space as a resource and
as a matter of design, and both follow a scheme of oppositions, albeit the way
they view these oppositions is completely different. Global limits to growth are
contrasted with boundless spaces filled with possibilities for all humankind, and
resource constraints are contrasted with an excess of matter. These oppositions
coalesce around powerful visions of undiscovered, utterly unknown landscapes
as they are projected in the macro- and nanocosm. This holds in particular for
the visual and conceptual opposition of spatial limitation (*Be-grenzung*) and of
spatial openness (*Ent-grenzung*) as guiding ideals of not only the nanoworld
but also the green world of sustainable development. Both claims – for spatial
limitation as well as for spatial openness – are continuously in play and are constantly being re-evaluated and renegotiated in respect to societal, scientific and
political debates. It seems that such visions of space make it possible to bridge an
otherwise irreconcilable gap between the politics of risk reduction and technoscientific experimentation with its hypothetical sense of promise and possibility.
Thus the spatial implications of sustainable development and green nanotechnology are of crucial importance for understanding both discourses – and, of
course, for seeing both as ecotechnologies.

11 ON THE PLEASURES OF ECOTECHNOLOGY

There is something between us better than love: a complicity.[1]

Marguerite Yourcenar

In the previous section, we saw how the programme of 'green nanotechnology' appears to bring together two strands that have commonly been seen as contradictory: on the one hand 'green' represents strategies of preservation by referring to the lawful limits to growth – in short, by saying 'no'. On the other hand there is the boundless space of 'nano' that is full of discursive and technical possibility and which elicits the pleasure of saying 'yes'. Looking for a moment at this 'no-but-yes' situation, we can perhaps better understand the seductive power of a term like 'green nanotechnology'. The promise of 'green' consists in part in its reference to the historical success of the environmentalist discourse, which seems to offer the possibility of saying no and therefore of bringing a limiting factor into play. However, as soon as it is brought into play, the notion of (unlimited) possibility introduced with newly emerging technologies – be it cabin ecology, nanotechnology or synthetic biology – draws us beyond these limits into a realm of technological opportunity. This difficulty of maintaining a 'no' while being pulled towards a 'yes' has also been given voice within environmentalist debates. In their provocative book *Break Through: From the Death of Environmentalism to the Politics of Possibility* (2006), Nordhaus and Shellenberger call on fellow environmentalists:

> to replace their doomsday discourse by an imaginative, aspirational, and future-oriented one ... We should see in hubris not solely what is negative and destructive but also what is positive and creative: the aspiration to imagine new realities, create new values, and reach new heights of human possibility.[2]

They ask explicitly 'how can [we] get from No-sayers to Yes-sayers? From pessimistic stories to optimistic stories?' At least some of their readers chime in: 'Nordhaus and Shellenberger are right. The Industrial Age gave us an environmentalism of limits and a politics of "no". The Creative Age requires a politics and culture of "yes"'.[3]

There appear to be two unusual factors operating here: first, the polemic against doomsday storytellers and negativity comes from the environmentalist camp itself and, second, the authors are seeking to establish a new eco-narrative, one that embraces technological innovation and reaches out toward a culture and politics of removing or exceeding limits. This breaks with frequent narratives about 'green' or 'eco' in the sense of taken-for-granted theories or so-called 'folk theories' that express a form of expectation 'based in some experience, but not necessarily systematically checked'.[4] Such folk theories are widely used and can be considered to be almost synonymous with 'off-the-peg' prejudices, that is, with the taken-for-granted stock of customs and cultural practices in a society. Despite their non-scientific nature, they are regarded as providing orientation for future thought and action.

For the sake of simplicity, 'eco' and 'nano' can be described as two folk theories in which the latter stands for any and every emerging innovation-driven technology. 'Eco' is usually seen as a preserving and conserving technology that draws on a vocabulary and various models of limitations and that keeps us within our 'limits to growth' or 'carrying capacity'; it addresses the need for a careful or even conservative stewardship of available resources.

The folk theory 'nano', by contrast, is usually seen as rather different in that its associated technological visions almost literally open up spaces and create endless possibilities that take us beyond our current limits. It is by no means obvious, therefore, that both nanotechnology and ecotechnology should come to signify the idea of sustainability. They come from different 'ballparks' – from innovation and conservation – and their stories refer to two opposing 'folk theories' that represent different relationships towards technology. If we do not wish to assume that green nanotechnology is merely a programme aimed at manipulating people and betraying their trust in order to secure economic success for nanotechnology per se, then we might want to work with a different hypothesis, namely, that the widespread label 'green' and the yes-saying of 'green cultures' simultaneously entails a change in folk theory, one that embodies a 'no-but-yes' and promotes the vision of sustainable innovation and the creation of win-win situations.

The Moral Economy of Hedonism

Sustainable development in general – and green nanotechnology or green biotechnology in particular – might be the most irresistible of all the currently available games that exude a seductive pull. What appears oxymoronic at first sight is apparently resolved when this 'no-but-yes' perspective of sustainable innovation is adopted in order to look at the development of green nanotechnology and sustainable population growth and economic growth with open-minded curiosity. All this appears possible if new spaces and new resources can be accessed, if limits exist only in relation to technology development, and if win-win situations can arise for technology, for the economy and for the environment. The seduction consists in the promise that, if everyone comes together to say 'yes', these win-win situations become possible and will indeed be realized.[5] The flipside to the generality of this promise implies an indeterminacy of meaning and a conceptual emptiness in both nanotechnology and sustainable development alike. Phrases such as 'Responsible Nanotechnology: Green Nanotechnology', 'Green Nanotechnology: it's easier than you think', or 'Nanotechnology provides "green" path to environmentally sustainable economy' are ubiquitous.[6] Just as in sustainability discourse, the 'green' in green nanotechnology refers to imperative necessities situated at the intersection of economic and environmental benefits while simultaneously retaining the technoscientific openness of the space of possibilities. Actors with radically different values and interests become caught up in a game that already seems to be paying off: when you advocate ecological sustainability you can become committed to economic sustainability, and *vice versa*.[7]

What do stories about the conquest of space, about win-win situations and the 'no' that is followed by a 'yes' tell us about the moral economy of ecotechnologies and, in particular, about sustainable development? The moral economy in question is that of hedonism, a regime of pleasure and the denial of limits.[8]

The first strategy of seduction that becomes apparent is that of a language game in which 'concepts mean whatever anyone takes them to mean'. When, for example, social conservatives, business leaders and environmentalists associate different things with the term 'sustainability', this does not imply that the proper meaning of 'sustainable development of nanotechnology' awaits determination. Rather, when using the term we please one another by denying that its meaning needs to be contested. The different associations do not lead to conflict because there is no perceived need to settle them, and there is no perceived need to settle them because the term is intended to mean only what any individual may take it to mean, rather than being parochial or conceptually pure. We might call this the pleasure of *Entgrenzung*, that is, an ecstatic transcendence of limits or a breaking down of the barriers that separate us. More generally, this *Entgrenzung* is at work in the sense that the conflation of concepts and meanings relieves us of all pressure to situate ourselves in this context and to take responsibility. In consequence, nobody feels any obligation to distinguish between different notions of technology or of efficiency or to subscribe to certain tasks (such as monitoring).

Thus, if 'we' (those of us who feel involved in these debates in a reflexive way) are not to lose credibility, we must claim that there is consistency between, for example, economic and ecological sustainability. And this claim of consistency itself provides pleasure because it affords an unlikely and unexpectedly harmonious alliance. At the same time, however, it constricts the meaning of 'ecological (or economic or social) sustainability' and thereby reduces the space of possibility, which may in turn be construed as spoiling the game (that prevents innovation and progress).

A second strategy of seduction follows the moral imperative of 'change for the better': when a second chance arises to 'get things right this time', a bond of hope-filled pleasure is forged by embracing this chance. In the wake of the so-called 'GMO disaster' (which is generally attributed to a lack of sensitivity towards consumer sovereignty and to the arrogance of leading figures in science and governance), nanotechnological innovation is being presented as an invitation to everyone to support its responsible development along with its concomitant promise of sustainability. And so we all join together to 'do the right thing', inevitably feeling good about ourselves and basking in the exhilarating pleasure of this celebration of civic virtue. As mentioned above, however, what this simultaneously does is eliminate the insignificantly small word 'no' from what had appeared to be an open discursive space.

A third strategy for mutual seduction results from the unboundedness of technical potential with its endless lists of benefits, its claims of 'all is possible' and of 'global abundance', with the whole of humanity as its beneficiary. Thus all are invited to expand imagination for the potential benefits of nanotechnology, and all proposals are poured indiscriminately into the common pool of possibilities. Rather than critique each claim on the basis of, say, its lack of technical feasibility, its societal exigency or of political justice as to the diversication of risk, the community of actors, once dragged into the game, welcomes them all. This encourages a kind of collective narcissism, with all involved partners immobilized by staring into the dazzling light of unlimited possibilities. The pleasure derived from this might be described as the 'hedonism of possibility'.

The Power of Seduction

All these strategies initiate a game of seduction[9] among actors involved in the game 'green nanotechnology'. Out of that pleasure-filled game emerge pressing commitments that take over the space of possibility. Without having been asked, our societies are, at this point in time, committed to a sustainable, European, efficient development of nanotechnology that is full of unspecified promise, such that these various values cannot be questioned or contested but must mean what we can all take them to mean. Playing with possibilities thus gives way to ineluctable necessities, as our mutual seduction invests empty concepts with the power to hold our conflicts and differences at bay. By being open to interpretation and at the same time remaining impervious to interpretation, these terms are capable of generating consensus among those who use them, imitating one another in a use that does not, however, presuppose a sharing of meanings. The persistence of this vacuous consensus proves powerful in that it forecloses a radical 'no'; it forces research and development into the straitjacket of 'efficiency' and 'high-tech innovation', and it discourages the articulation of non-stereotypical visions and even the assertion of special interests (those of workers, of different generations, of people with disabilities etc.).

Seduction entangles us with one another as we heed a call towards commonalities and enter a richly textured discourse, finding ourselves in more or less voluntary association with others who are playing the same game. And the very fact of this association around a practice and a game – around the occasion provided by a word devoid of determinate meaning – creates the pleasurable illusion that we share a common ground from which everyone can pursue their goals without compromising those of the others. This is the way in which positioning and politics drop away and power creeps in, namely, the enchanting power of pleasure:[10] a euphoric agreement on an inventory of symbols, words and rituals – but only in reference to the future. We end up mutually enchanted, committing ourselves to something we cannot fully survey and agreeing to more than we can possibly mean.

These and other pleasures are at work in the unlikely alliance between business and environmentalism, in the fusion of interests in win-win situations, in the displacement of all conceivable problems by technical solutions, in the seemingly endless and playful diversity of images and objects in apparently unlimited space, in the *jouissance*, exhilaration and enthusiasm experienced when forging possible futures, and also in the exploratory spirit of experimentation that has escaped the narrow confines of the laboratory. All this opens up seemingly irresistible possibilities of gaming and tinkering in a world that seems to be ready to hand and that appears to be free of any traces of history or limiting representations: it is a world that simultaneously offers facts and fictions and that invites us to enter it and move around in it, probing it extensively and passionately. Thus the fixtures and fittings of the ecotechnological world are constitutively ambiguous ones and need to be adjusted again and again; this is a world in which the 'oikos' is in a permanent mode of experimentation – all of which provides a taste of the likewise pleasurable and precarious work to be done in the context of a knowledge society contract.

12 CONDUCTING A SOCIAL EXPERIMENT: 'BUILDING WITH NATURE'

In the following the provisional conclusion reached at the end of the previous chapter will be scrutinized: what does it mean to talk of the ecotechnological 'oikos' as being subject to a permanent societal process of probing and tinkering involving a similarly precarious and pleasure-filled drive towards innovation? Probing and tinkering in a social context can be fruitfully described in terms of a social experiment as long as a few minimal conditions are fulfilled. Two of them may be that, first, the problems to be solved are framed in a way that includes an intersubjective process of hypothesis building and, second, that the – more or less passive – participation of those stakeholders identified as having an interest in the matter at hand is guaranteed. It is most likely that social experimentation does not necessarily need to be an activity that is instigated by social (or political or economic) scientists or involves (natural) scientists. To apply the semantics of social experimentation in a beneficial way, it is sufficient to identify a situation that is perceived to be a problem in the real world and implies the need for social innovation. This may be the introduction of a new technology (broadcasting, mobile phone), the planning or/and restoration of an urban district or a large-scale technical facility (an embankment dam, an urban green belt, a power plant), or the introduction of a new regulation (the European REACH legislation, American green chemistry regulations, Chinese environmental legislation) or of new consumer products (GM foods). In a nutshell, virtually every societal and social activity can be fruitfully conceptualized as a social experiment if this is believed to be a useful language game.

Social experimentation can be regarded as a type of field experimentation. Accordingly, it suggests itself in the context of this study to call to mind the three criteria for field experimentation as proposed in the section 'Dirty Places of Experimentation' in Chapter 6 above. To recall them briefly, these criteria were described as follows: first, identification of the relevant boundaries and how they can be conceptualized adequately; second, decision about what kinds of sense data are to be collected with respect to the hypothesis; and third, examination of the types of instruments and expertise needed to perform measurements.

Implementing Values in Technological Artefacts

A number of conceptions have been offered above to describe different modes of technological materialization, what they contribute to societal transformation, and how particular social values exert a formative influence. Another important factor that plays a decisive role in this game of societal power relations is the question of what designers contribute to the technological transformation of society. What the conceptions selected here have in common is that they are committed to the idea of democratizing technological cultures – broadly speaking, to bring both social and technoscientific influences under democratic control.

At the end of Part II, real-world experiments were introduced as one type of social experimentation, with particular emphasis being placed on recursive learning processes as a typical feature of innovation-oriented projects. Another type of social experimentation involves looking at transformative processes in technological materialization in terms of the 'self-reproducing potential' of socio-technical arrangements.[1] Basically, this theory says that once social values are settled in a certain technological design, they will also contribute to the development and formation of socio-technical practices. It is these socio-technical practices developed conjointly by the partners involved in a project that afford new conceptual and material solutions and thus new technology design. In this way, the new values and visions agreed upon can also be integrated into the design and construction of the model and, eventually, of a park or of a building.

It is this theory of the joint formation of socio-technical practices that latently informs the following discussion of the inventive and mediatory role played by a three-dimensional analogue model in bringing into being a novel artefact – an ecologically fully functional bypass channel – in the context of landscape architecture work undertaken in connection with a newly constructed hydropower plant.

'Building with Nature': Building with Rather than against the River Rhine and its Neighbours

The following story is about two technoscientific objects, the hydropower plant Rheinfelden and its bypass channel, both of which embody the latest success stories in engineering, the former in hydromechanics and hydroelectric turbine technology and the latter in landscape architecture and ecology. Both of these objects are outstanding high-tech constructions and they both stand for cutting-edge research and innovative engineering solutions: each is a superlative in its own right. Both of these technoscientific objects are based on a form of knowledge production that uses scientific representations to make things work; in both cases there is a rich interplay of 'representation and intervention' in terms of research technologies, instrumentation, experimental techniques and modelling practices. Particularly salient for the following analysis is a philosophically interesting feature of technoscience that is the application of theories as tools for constructing models of particular phenomena.

The bypass channel is the outcome of an extraordinarily successful social experiment in which a so-called monitoring commission played an especially important role. Another important element in this social experiment was a 3D model, the so-called demonstration model (*Anschauungsmodell*). This model, which accompanied the whole design and planning process, came into existence even before the building application was approved.

Setting the aerial photo and the 3D model photo side by side (see Figure 12.1 and Figure 12.2), a striking similarity appears between the recent real-world object – the artificial running water course as it is – and the *vision* of the real-world object as it is represented by the demonstration model. It is a central thesis of this investigation that this similarity was achieved by a bi-directional process of tuning and adjustment, of probing and tinkering with the model through material interventions.

In a first step the model was designed and built according to the visions, theories and data provided by the landscape designer and engineers. It was subsequently modified in order to respond to material and socio-political resistance in the ongoing planning process; finally, the artificial running water course was built according to the modifications presented in the model, embodying so to speak the discursive processes visualized therein – chiefly the conception of 'building with nature'. The 3D model successively incorporated the ongoing debates on the project and also helped to visualize and clarify them – 'let's go to the model' was a commonly used expression during the planning and construction phase.

Figure 12.2: Aerial photography of the Higher Rhine near Rheinfelden, where recently the construction of the hydropower plant and the so-called 'near-natural bypass channel' was finished. The aerial photography was produced by Luftaufnahmen Meyer, Rheinfelden, and reproduced courtesy of Energiedienst, Rheinfelden.

Figure 12.1: Photography of the demonstration model 1:200, showing the 'near-natural bypass' and parts of the hydropower plant above water level. Reproduced courtesy of Energiedienst, Rheinfelden. This photo of the demonstration model appeared first on the front page of the environmental impact study in 1994, and it was modified repeatedly until 2008.

Genesis of a New Technological Materialization in the River Rhine

The new power station is a run-of-river power plant whereas the old hydropower station was a diversion-channel type. This basically means that the hydroelectric turbines are arranged horizontally to the direction of flow and that the power plant is built across the river – which in this case also means across the Swiss-German border that runs across the middle of the river. It is the near-natural bypass channel (situated on the German side) which will be focused on mainly in the following discussion. This constructed 'piece of nature'[2] is the largest fish pass facility of this type in Central Europe, as it says in the prospectus of Energiedienst, the company operating the hydropower plant.[3] The construction licence for the project was awarded in 1998, and in the same year a monitoring commission (*Begleitkommission*) was established that was integrated in various ways in the early phase of construction and during the ongoing design process. The construction work began just five years later and was officially completed in 2012, when the bypass channel was inaugurated. At the time, the channel had already been conceptually transformed into a semi-natural running watercourse (*naturnahes Fliessgewässer*), as it was called in the media. However, because the planning phase of the new hydropower plant had already started in 1989 with the application for a new licence, the lifetime of the project extended over roughly a quarter of a century in all.

The monitoring commission was involved in the design activities as early as 1998 until completion of the project in 2012, when the finished bypass channel was approved and accepted. The commission was composed of environmental organizations, angling clubs, municipalities and other stakeholders, all of them participating in a step-by-step procedure in the design and planning process. Overall, this was perceived as a confidence-building measure. The significance of the commission is confirmed by the leading engineers, both of whom felt that values such as trust and justice were of crucial importance for the realization of the whole project.[4] Similarly, both of them confirmed that the establishment of the monitoring commission was an important component in the socio-political implementation of this technological project. It is interesting to note that the process of deliberation is still not over; there is an ongoing discussion on both sides of the river (in the Swiss as well as the German municipalities and NGOs) about what kind of nature people want to experience and how this should be managed. This set of concerns is sometimes translated into questions about what kind of nature might be good for non-human stakeholders such as fish, birds or beavers.

However, it is quite obvious that these kinds of questions differ in character from the deliberative processes that took place during planning and construction, when the engagement of different stakeholders in the ongoing design process of the power plant itself as well as of the bypass channel was a dominant feature of the whole project. Participation and engagement were guaranteed institutionally by the monitoring commission, which was able to exert a direct influence on decision making during the construction of both facilities. Thus the commission influenced not only the design and materialization of the power plant itself but also the conception and planning of the compensation measures. However, this materializing power also worked in the other direction: through its involvement in these large-scale technoscientific projects, the commission's conceptions about nature and technology also changed over time, simultaneously influencing the way its members developed and discussed values and norms associated with the projects.

This observation fits well with STS approaches that place emphasis on the material stabilization of values and norms through technology. Such approaches hold that technologies influence the actions and thinking of those who are involved in developing and using them and that this in turn affects which values and norms are realized and which are not. Different technological cultures show different politics in the appropriation of technology design and have developed different sites and modes of decision making in the process. They might differ, for example, according to what kind of democratic model is involved, what the local political structures are, what the role of the designers (including architects and engineers) is in these constellations – and whether there is a preference for a culture of experts or whether participation and the development of common values is more important.

This is precisely the point at which the expression – or, perhaps more appropriately, the *tool* – 'building with nature' comes into our case study.[5] 'Building with nature' – *Bauen mit der Natur* – was the overall concept that ran like a thread through the project 'Reconstruction of the Rheinfelden Hydropower Plant'. 'Building with nature' became a crucial tool when it came to planning the near-natural bypass channel and to the subsequent construction of the analogue 3D model. It was mainly the latter which served to trigger the development of common values and norms with and within the monitoring commission and to bring down imaginaries about 'nature and technology' to be incorporated in the project. To make this happen, that is, to materialize the values and norms associated with 'building with nature', the deliberation and development of the socio-technical practices happened in close contact with the 3D model. 'Let's go to the model' became an important part of the socio-technical practices that served to settle social, functional or aesthetic claims in the design process. And due to the fact that the design process developed over more than two decades, it was necessary constantly to adjust the socio-technical practices to the changing conditions in the physical world outside and in the social world between all those involved in the project. In other words, people were tinkering with concepts, tools and theories in order ultimately to engage in consensus-based decision making. It is this social experiment that is reflected in the biography of the model.

Biography of the Demonstration Model

The model was designed and manufactured by Christian Oehrli from the Oehrli-Fricker architectural model making company in Zurich. It was originally built in 1991 and was last modified and actively used in 2008. Today it is laid out in the machine hall, exposed to the permanent noise and vibrations of the turbines. The scale of the model is 1:200 and it measures about 1.70m in height and 2.30m in length. It is built of a host of different materials.

The base of the model is made of wooden frames, while the shape of the landscape is constructed from plywood ('air-ply'). Most of the houses are made of polyurethane, and parts of the so-called 'Gwild' – the cliff-like structure in the middle of the Rhine – are modelled using foamed acrylic glass. Water and its flowing movement have been drawn using colouring pencils and have been treated with a finish. From the very beginning the model was built in two parts in order to facilitate its transportation.

Like the material, the sources of information that came to be implemented into the model were of manifold origin: there were topographical and hydrological data, raw data from the buildings and photographs of the landscape, the urban environment and different sites on the river. Not only was this information presented in different types of representations and media, it already contained a great deal of processed information from other models or maps. The model served, in a sense, as a data transformation tool, gathering and assembling the vast amount of information available and transforming it into a form that could be readily grasped; it proved to be good enough for experts and non-experts alike. The experts recognized their theories and models of different scales, while the non-experts were able to identify at a glance the landscape and the shape of the new technical structures and could bring in their own local experience and knowledge straight away.

It is the engineers who talked about the model in terms of a demonstration model, and still do so. This may indeed be a suitable term, provided 'demonstration' is understood in an active sense – in other words, not as a teaching model but rather as a visualization that incorporates within itself the conditions of the cognitive and technological production of seeing (which is most often not the case when the engineers are talking about the model). However, understanding the model as a transformation tool underlines the active aspect of the model mentioned above in relation to its mediating role between engineers, experts and the monitoring commission. In addition, the model was used in presentations given to other stakeholders and to invite a broader public to discuss the project.

To do so, the model was transported to the places where the public debates were held or the expert groups were located. Given this mobility, it seems more appropriate to talk of the model in terms of a nomadic object that brought people together and gathered information from other models and representations by focusing the discourse on the nature of the bypass channel.

Thus the model provided an opportunity – along with a certain material malleability – to bring in new ideas and theories and to probe these ideas in situ, albeit at a scale of 1:200. The openness to modifications is immediately visible in the model in that it is divided into several separate sections that can be lifted out, thereby facilitating the process of modification and adaption to shifting conceptions and values. This can also be seen as a material contribution towards the self-reproducing potential of socio-technical arrangements, affording a new design for the jointly developed new solution.

So far, we have taken a closer look at the material conditions of the model and their implications for the socio-technical arrangements associated with the project. In a next step we will deconstruct the model to an extent, exploring the sources for its construction and the theories, models and maps involved. At the same time this will bring us closer to answering the question how the principle 'building with nature' is involved in technological materialization, whatever form it may take.

From Demonstration to Transformation Model

It has already been mentioned that the model operated as a data transformation tool. Accordingly, in the following we will try to figure out some of the data types that in turn make up the Oehrli model and how the principle 'building with nature' is translated into a technological materialization by this process of transformation.[6]

A near-natural watercourse should ideally comply with the following stipulations: to operate as a fish pass and offer running water habitats, it is equipped with a river bed, to provide a high variety of bottom and flow structures induced by sequences of riffles and pools, and single gravel islands, among other features. The function of these structures is to establish habitat connectivity between the upstream and the downstream section of the power station but also to provide suitable habitats for so-called rheophilic species, animals with a preference for running water. According to the landscape designer, the bypass channel as a whole is like a mountain river, offering 'good integration into the landscape' and 'aesthetic value'.[7]

Aesthetic value and knowledge about the 'natural' character of a mountain river is drawn mainly from a number of real-world models that are characterized by 'natural development'; one example named is a section of the river Rhine close to Breisach. These images are placed virtually on top of each other, together providing the ideal type of a mountain river. This ideal type was tested by applying it to another project design of a bypass channel on the river Aare (shown in the upper left-hand corner) which was much smaller than the one in Rheinfelden. The Aare model was the specific, tangible 1:1 model that materialized the tool 'building with nature' and also confirmed it. The model was also integrated in the Oehrli model and eventually contributed to the adjustment of the socio-technical arrangements in Rheinfelden.

Another example is the so-called physical model produced by the Hydrological Institute in Karlsruhe. This 1:50 model was used not only to determine the necessary flow rate of one of the fishways but also afforded the insight that the 2D plan and its associated calculations were erroneous. It was only the physical model that was able to provide the right solution – that is, a functional solution to the situation. This was also integrated into the 3D model and helped to optimize the ideal model of a near-natural running watercourse.

What does 'Building with Nature' Mean, Then, in this Project?

The Rheinfelden project seems to follow a principle of mimesis, understood as a principle of transgression – that is, not simply in the sense of mere copying but as a process of creation out of which something new emerges. Building with nature here has something to do with the notion of 'using nature to go beyond nature'; the idea seems to be to outdo nature in its given form.[8] What is manifested here is a technical and aesthetic construction of a specific piece of nature *qua* model. This new nature is constructed *through* the model of an ideal landscape and is built in this particular place and at this particular point in time – better than nature could ever have done so itself. This newly built piece of nature is an ideal type that unifies within itself an array of real places and times all at once and yet is a material reality, recognizable as a landscape. The gathering of conditions in particular places at particular times is ultimately what brings forth the ideal typically built piece of nature. The process of gathering is revealed in the development of models and in the links through which these various model types are connected. The manifold connections between the models are created and secured, for example, by the practice of construction, by the materiality of the models, by aesthetic properties and by the inscription of epistemic values. The reading of 'building with nature' in this project is all about innovation: it is the construction of a new landscape using planning techniques and design elements; technological development, scientific research and social innovation work hand in hand with one another. This work is performed by the 3D model – the so-called demonstration model – that transforms the many theories, models and practices it embodies into a recalcitrant object and, in doing so, contributes to the material stabilization of values and norms through technology.

13 POLITICAL ECONOMY OF EXPERIMENTS

Are there actually conceptual models in science that attempt to calculate the intangible and objectionable dimension of time (measurable as it is only by means of secondary phenomena such as change and motion) out of physics?[1]

Wolfgang Herrndorf

When researchers account (literally) for physical processes such as the management of matter, energy, masses and of space and place, they follow certain housekeeping principles. To do so, they have developed numerous instruments and techniques of intervention as well as various ways of representing the results. All these activities are essentially economic activities: weighing, recording, measuring, metering, comparing, calculating, consolidating and, finally, accounting for what all these activities have provided in terms of theoretical knowledge and experimental practices. What makes this a *political*, rather than a moral or cognitive economy,[2] is the fact that a decision is to be made about how the housekeeping is to be managed and what is to be its benefit (theories or products, discovery or innovation, fundamental or applied knowledge, etc.). Essentially there are two modes of housekeeping: one is an economy that follows a logic of adaption to limits while the other is an economy that strives for conquest and thus seeks to transgress limits. Given the foregoing discussion on the practices of experiments, it is apparent that these are decisions of eminent political relevance. The case studies provided in Part IV give an impressive account of this, be it the promise of 'green technologies', a sustainable development that is offered as a substitute to conservation, or the construction of ecosystems that are better than any natural ecosystem could ever be in terms of efficiency and local adaptivity. Further examples not included in this book come to the mind, such as the search for inherently benign technologies that are safe by design and the idea of enhancing material nature, be it by using biotechnologies to turn a genetically rich collection of local crop plants into an industrially commodified product like golden rize[3] or by using nanotechnology to turn dead matter into smart material. Again, other examples are approaches in climate change research that seek to develop social and technical tools to prepare 'us' for cultural transformation while 'we' are going to adapt to climatic change,[4] or activities in the ecotechnological design of nature, as in the previous chapter's discussion of 'building with nature' as a social experiment.

Accordingly, the intention of this section is to reveal the political economy of experiments when they are put to use in either a scientific or a technoscientific context.[5] The aim here is more than simply to situate scientific activity within the political economy of a society or to investigate the role of science and technology in economic growth and the relations between science, technology, the state and capital.[6] More specifically, the idea is to develop a philosophical account of housekeeping itself. This will be done here by observing how researchers manage matter or energy, how they 'fabricate' their experiments, as Hannah Arendt so tellingly characterized the activity of researchers when engaging with the material world. How do researchers negotiate space, surface area and place, and how do they approach their research problems? Do they accommodate themselves to limits and constraints or do they seek to overcome such limits?

This approach of an economy of experiments may bring to light the underlying assumption in much contemporary research practice, namely, that of an unlimited technoscientific world of abundance and excess which challenges the received certainty of a limited world that rests firmly and solidly on physical laws of conservation and on a conception of space as a radically limited resource, as discussed in Chapter 10, particularly the section on cabin ecology. This juxtaposition of a limited and an abundant world in terms of ontological and epistemic assumptions is far from new. In the context of a political economy of society, it was already present in the work of philosophers and sociologists such as Werner Sombart and Georges Bataille during the first half of the twentieth century. They contrasted two political economies around the notions of a limited world defined by conservation and an unlimited world defined by luxury, excess and abundance. In doing so, Sombart and Bataille both recognized the central role of science and technology in modern economies.

Indeed, Bataille goes as far as substantiating his economic conception by referring to scientific models including, most significantly, the concept of biosphere: 'The terrestrial sphere (to be exact, the *biosphere**), which corresponds to the space available to life, is the only real limit'.[7] Because Earth is exposed to a permanent input of solar radiation, there is always an excess of energy. As long as living organisms grow and proliferate and solar energy is properly absorbed, excess is minimal. However, this situation changes once the limit of growth is achieved: 'life ... enters into effervescence: Without exploding, its extreme exuberance pours out in a movement always bordering on explosion'.[8] In a fully realized biosphere 'there is generally no growth but only a luxurious squandering of energy in every form'.[9] While humanity has successfully extended the limits to growth by investing in labour and technology, it also has immense power 'to consume the excess of energy intensely and luxuriously'.[10] These are the assumptions which Bataille develops further when he regards the role of wealth and excess. In his so-called general economy the 'limit to growth' opens up a world of excess, which is an unavoidable aspect of all production and all transformations of matter. By contrast, a restricted or special economy takes 'limits to growth' as a challenge to live productively within one's means and to gain surplus value from processing and reprocessing finite resources. But although he drew on scientific ideas to establish his conception of general economics and despite distinguishing the scientific, restricted economy of the economists from his avowedly non-scientific general economy, Bataille did not explicitly articulate a contrast between different ways of conceiving and exploring the world, between a general and a restricted political economy of science.

In the following, Bataille's distinction between a restricted economics and a general economics is expanded to develop the philosophical notion of a political economy of science and a political economy of technoscience. The political economy of science is constituted by conservation laws and is therefore implicitly committed to an idea of limits to growth.[11] In contrast, general economics can be identified with the technosciences, which appear to adopt a principle of non-conservation, innovation or infinite renewal – as exemplified, for instance, by the ambition to expand resources like 'space' (cabin ecology) or matter (biotechnology).[12] The classical conceptions of a conservative science appear in each of the disciplines of physics, chemistry and ecology. Likewise the tendencies to exceed those limits characterize nano- as well as ecotechnologies. The interest of a political economy lies in the different treatment of space, matter and energy that is reflected in different research practices. These practices are not neutral; they are of eminent political significance. On the level of the norms of representation and ideals of production that govern scientific and technoscientific research, these practices condition political choices. With respect to the restoration of nature or to global warming, this entails the choice between mitigation or adaptation and expansion of capacity.

Economic Principles in Knowledge Production: Balancing or Exceeding?

Conservation principles are as old as science itself. However, Antoine Lavoisier's formulation of the conservation of matter holds a special place because it provides an obvious case of scientific reason acting as a lawgiver to nature: 'in all the operations of art and nature, nothing is created ... the quality and quantity of the elements remain precisely the same and nothing takes place beyond changes and modifications in the combination of these elements'.[13] Lavoisier continues by pointing out that this principle usefully serves as a standard for chemical experimentation and hypothesis formation: every experiment must submit to the housekeeping authority of the scale that demands a complete account of the entire quantity of matter before and after the experiment. Only such experiments and only those hypotheses that meet this demand are admissible.[14] And just like the Lavoisian chemist, economic theorists represent the world by way of accountancy: a restricted or special economics, from Adam Smith and Karl Marx to John Maynard Keynes, considers how wealth becomes concentrated and distributed; it looks at the circulation of goods and currencies, at the balancing of cost and price and of demand and supply. The creation of wealth and even of 'surplus value' is accounted for in terms of the extraction of material and human resources.

This reliance of certain experimental practices on conservation principles and therefore their utilization in scientific knowledge[15] was challenged by Georges Bataille. He put forward, as an antonymic concept, a general and unrestricted economics that celebrates excess and waste and interprets them as gifts of the sun.[16] This creative aspect of excess and waste appears in the thought of economist Joseph Schumpeter, who refers explicitly to Sombart (and Nietzsche) when he proposes that 'creative destruction' is a basic economic process. However, Bataille did not simply propose a different theory but a different and explicitly non-scientific form of knowledge production. Since he understood the role and function of restrictive conservation principles as conditions for scientific knowledge, he pointed out that restricted and general economics are accompanied by two kinds of knowledge.

The scientific knowledge of special economics is born of an anxious concern for particular facts and is characterized in terms of an all-encompassing calculation, as everything needs to be accounted for. According to Bataille, it 'merely generalizes isolated situations' and 'does not take into consideration a play of energy that is not limited by any particular end'; furthermore, it does not consider 'the play of living matter in general, involved in the movement of light of which it is the result'.[17] This play of energy is excessive in that it exceeds the accountants' balance and produces a surplus. It is therefore not the subject of conventional scientific knowledge, and Bataille hints accordingly that he wants to add to the wealth of knowledge even as he must fail at being scientific:[18] a general economics that takes as its model the sun's gift of energy to the Earth does not account for the creation of wealth or of knowledge as a mere redistribution or re-presentation of available resources but views all wealth production (including that of knowledge) as a sign of abundance, excess and general surplus – as something that must be squandered and cannot be earned. Thus when Bataille pursues his general economics he follows the movement of energy from geophysics through biology into society, not in terms of income and matching expenditure but in terms of excess and destruction.

On Bataille's terms, then, scientific knowledge like that of Lavoisier depends on the counterfactual construction of special and limited principles of conservation within the more general movement of unlimited energy. Within the thermodynamically open system 'Earth', chemical laboratories are established as closed systems for the sake of the scientific presentation and representation of isolated facts. Since the excessive movement of a general economy of nature and society tends to undermine the creation of isolated closed systems and thereby human interests in intellectual mastery and technical control, the special sciences satisfy those interests for better or worse.

It is now possible to see in what sense Bataille's general economy is 'non-conservative', namely, in the sense in which there are non-Euclidean geometries that include Euclidean geometry as a special case, or in the sense in which Gaston Bachelard speaks of non-Lavoisian chemistry.[19] Just as non-Lavoisian chemistry follows the material processes of purification and experimental isolation that yield the kinds of substances and representations which then allow us to see nothing but recombinations of elements, so special economics appears as a limited case of a general economics that focuses on the way in which a special economy is constructed.[20] From the point of view of general economics or non-Lavoisian chemistry, conservation principles ensure a constancy of nature as a necessary prerequisite to allow for scientific inference and representation. With this conception of conservation principles in mind, one will start seeing them in all efforts to represent scientifically the features and causal processes of the world as they are expressed in the venerable laws of 'nothing can come from nothing' and 'nature makes no leaps'. The scientific world is committed to the conservation of mass and energy, of charge and angular momentum; there is uniformitarianism or actualism and there are the so-called inductive principles which posit that the future will be like the past and that nature does not change. All of these speak of the world as a limited whole in which nothing is created and where all change is a redistribution of what is already available. All these notions are introduced as prerequisites for scientific representation; they are representational norms that structure a domain of phenomena such that objective knowledge about it becomes possible.

As the complement to science, the technosciences surrender this supposition of a limited and balanced world – they acknowledge limits only to discover a world of excess and technical possibility within and beyond them. However, if technoscience is to science somewhat as Bataille's unrestricted economics of excess is to a special economics of limits, this does not amount to a scientific revolution or a paradigm shift in science where the new paradigm is non-conservative. If the argument so far is correct, there can be no such thing as a Kuhnian paradigm that is not constituted by one conservation principle or another. For the same reason, this shift does not involve a dispute about the standing of conservation laws: by definition, these are pretty much beyond dispute. Whenever technoscientists turn to the business of representation or explanation, they will be careful not to violate the conservation principles that serve as the representational norms in their community.

The claim that technosciences are non-conservative does not refer to agreements or disagreements about principles but to the idea of novelty and creativity, of an ontological difference that is implied in the making and building of things, in the acquisition of capabilities for the control of phenomena, in shaping or disclosing a new world. Technoscience stands for an embrace of the technological or constructive character of science. However, technoscience goes beyond these principles, seeing them as a constraint and thus as a challenge to probe or even transgress the implied limit to creativity and novelty.

Creative Experimentation while Simultaneously Obeying Nature?

On the account presented so far, controlling and managing the flow of energy and matter is a concern that is at the heart of science and technology in the modern world. For the scientific enterprise, the conservation principles ensure the constancy of nature and thus enable scientific inference and representation. And it is assumed that only this constancy and lawfulness of nature underwrites the technological enterprise. This conception of technology as applied science implies that technological ambitions and experimental creativity will always be constrained by a scientific worldview. Lavoisier's verdict on the limits of art and engineering and his view of experimentation emphasize that the scientist must literally surrender to the verdict of the experiment. Nature is invited into the laboratory as a witness who provides answers to our questions. Scientists, artists and engineers thus learn from nature not how to do or build things but rather which of their ideas are in accord with it.

This supposed impossibility to create genuine novelty also produces ambivalence. If nature can be used for scientific and technological purposes precisely because of its lawfulness, do we therefore have to surrender to the world as it is or can we still overcome nature and liberate ourselves from our natural condition – just as the conquest of outer space was seen as an attempt to leave our Earth-bound existence behind?[21] Bacon's conception of the experiment as a new style of innovative practice reflects this ambivalence. It expressed an experimental spirit that demonstrated in part its power to determine what is and what is not in accord with nature. At the same time, the experiment appeared as a technology for innovation, as a tool – a 'heuristic hoist'[22] – for transgressing given natural limits.

In the classical or conservative idiom of 'science and technology', the scientific assumption of a limited world sets limits for technology and experimental practice. Despite the noted ambivalence, it took a long time for the inverse relation to gain recognition: only in the fairly recent idiom of 'technoscience' does the unbounded creative potential of technology offer the expectation that the world, too, is unlimited. Accordingly, the idea of overcoming nature that is associated with creative experimental interventions comes to the fore. Instead of Lavoisier, it is now Francis Bacon who is claimed as a founding figure. Today's accounts of the Baconian experiment emphasize a spirit of creative experimentation that has gradually conquered modern societies by being adopted by artists, engineers, instrument makers and social reformers. All of them share a creative desire in designing machinery, creating artwork, exploring the globe or changing society.

Exceeding the Limits

In the following we take a closer look at this shift from the scientific conception of lawfulness and constancy as a limit on engineering to engineering practice as a model for the ability to exceed limits, including those that appear natural. This inversion is particularly evident in respect to the notion that there is limited space on Earth that constrains human civilization, including technology.

Economist Thomas Robert Malthus was convinced that man cannot transgress the absolute limit given by nature. In saying this he referred not only to limited space on Earth but also to other resources. In his famous *Essay on the Principle of Population, as it Affects the Future Improvement of Society*, published in 1798, he provides a clear and seemingly inescapable account that has not lost any of its power even today. Until at least the 1970s and possibly even beyond then, Malthus's population model fed into scientific ecology. It also influenced ecotechnologies like the cabin ecology of the 1950s and 1960s and the development of the biosphere in the 1980s. Today it serves as a basic assumption of ecological economics – an exemplar, to be sure, of what Bataille called special or restricted economics.[23] So what does the so-called Malthusian law say, and what makes it so particularly seductive for ecological concerns? It draws a causal connection between the growth of population, space as a limited resource and the availability of food production. In Malthus's logic, the scarcity of food resources is the absolute limit for societies; it is a scarcity that is implied by a lawful nature.

> I say, that the power of population is indefinitely greater than the power in the earth to produce subsistence for man. Population, when unchecked, increases in geometrical ratio. Subsistence increases only in an arithmetical ratio ... By that law of our nature which makes food necessary to the life of man, the effects of these two unequal powers must be kept equal ... Nature has scattered the seeds of life abroad with the most profuse and liberal hand. She has been comparatively sparing in the room, and the nourishment necessary to rear them. The germs of existence contained in this spot of earth, with ample food, and ample room to expand in, would fill millions of worlds in the course of a few thousand years. Necessity, that imperious all pervading

law of nature, restrains them within the prescribed bounds ... And the race of man cannot, by any efforts of reason, escape from it.[24]

This law of nature, according to Malthus, is the immutable condition for the economy and governs the relationship between humanity and nature.

> It accords with the most liberal spirit of philosophy, to suppose that no stone can fall, or a plant rise, without the immediate agency of divine power. But we know from experience, that these operations of what we call nature have been conducted almost invariably according to fixed laws. And since the world began, the causes of population and depopulation have probably been as constant as any of the laws of nature with which we are acquainted.[25]

Moreover, the scope of action for accommodating the 'great machine' society to nature is rather small because the lawful order of society is ultimately based on the force of human 'self-love' which is again given by nature. The Malthusian society cannot change the actual relations between the rich and the poor, and indeed it is not even disposed to imagine societal change.

> [A] society constituted according to the most beautiful form that imagination can conceive, with benevolence for its moving principle, instead of self-love ... would, from the inevitable laws of nature ... degenerate into a society, constructed upon a plan not essentially different from that which prevails in every known state at present; I mean, a society divided into a class of proprietors, and a class of laborers, and with self-love for the main-spring of the great machine.[26]

These somewhat 'anti-social' elements of Malthus's philosophy might disguise the originality of his economic model, which lies in the identification of nature as a resource for society and which still informs economic and ecological thinking today. The ground for this persistence is that nature is imagined as a given and unchangeable source that stands in opposition to the individuals existing in a society, including the trader and the economist. Societies have to accommodate themselves to this nature, and the limits of technology are imagined accordingly: nature cannot possibly be conceived in a technological manner. Malthusian nature, characterized by an unavoidable logic of power and balance, shares its fundamental assumptions with other conservation principles. Both the relationship between economy and nature and, above all, the economy of nature itself are construed as an inescapable necessity. An order is projected onto nature, and this order corresponds to the form of scientific laws and of economic processes alike.

Control and Balance outside the Laboratory

It is therefore hardly astonishing that the first and second laws of thermodynamics[27] have always played an important role in scientific ecology. 'The basic process ... is the transfer of energy from one part of the ecosystem to another',[28] wrote aquatic ecologist Raymond L. Lindeman when he first described a lake as an energetically open ecosystem consisting of biotic and abiotic components. Energy from the sun is accumulated in organisms (so-called producers) by means of photosynthesis. A portion of this energy is transferred via consumption to the next levels, but most of the energy is lost either by respiration or decomposition. This first description of the transfer mechanisms in an ecological system in terms of the laws of thermodynamics was further developed in ecosystem theory and thus became an important conceptual reference in ecological economics as a scientifically certified description of nature.

Lindeman's model was by no means the first one to conceptualize natural systems outside the laboratory as quantitatively recordable entities. As early as 1926 Vladimir Ivanovich Vernadsky published a paper, 'On Gaseous Exchange of the Earth's Crust', in which he treated geochemistry as a natural history of terrestrial chemical elements. This geochemical approach turned the whole globe into a scientific object, and it was from this that Vernadsky's concept of the biosphere[29] derived its heuristic power. The idea of control and balance is everywhere in evidence, since the objective of quantitatively describing the transfer of substances through a system is pursued by conceptualizing a biologically controlled flow of atoms in a specific geological site. 'All points oscillate around a certain fixed mean'[30] was one of Vernadsky's central statements which clearly expresses conservation principles.

Regulatory feedback mechanisms played an important role in Vernadsky's model: they structured conceptualizations of cyclical processes. In the 1940s this geochemical approach found its way into more general debates in systems analysis in the context of the famous Macy Conferences (1946–53). The inaugural meeting of the group was called 'Feedback Mechanisms and Circular Causal Systems in Biological and Social Systems'.[31] The very idea of a cyclical process and self-regulating feedback mechanisms constitutes a variant of conservation thinking. It serves as a norm of representation that constitutes scientific practice and also constitutes a specific scientific object, namely, a kind of system, including the ecosystem. Just like the post-Malthusian systems of agriculture, these systems can be more or less efficient in that they use the available space or the available energy more intensively. This intensification takes place strictly within the circulation of matter and energy: Malthus was 'proven wrong' only because he underestimated what intensification could do, but on this account he is still considered right in principle: there is a limit to intensification and this limit brings us up against Malthus's unyielding, unforgiving nature. Some of the participants of the Macy Conferences testify to this as they went on to develop what was later called general systems theory (GST).

Aside from neurophysiologist Ralph W. Gerard and ecologist George Evelyn Hutchinson, who was also an important supporter of Lindeman's thermodynamic ecosystem concept,[32] other major contributors to GST were biologist Ludwig von Bertalanffy, economist Kenneth E. Boulding, who in the 1960s advocated the concept of 'carrying capacity', and biomathematician Anatol Rapoport, whose thermodynamic models influenced ecological economics in the 1970s. Hutchinson's seminal paper on 'Circular Causal Systems in Ecology' shows clearly that his 'systems' are scientific objects that are to be studied and represented by scientific ecology. He developed an ecological theory using cybernetic terms such as feedback mechanisms and circular causality, arguing that, within certain boundaries, ecosystems are 'self-correcting' by means of 'circular causal paths'. The assumption that these regulating feedback systems constitute ecological theories forms the basis of both his biogeochemical and biodemographic approach.[33] Abiotic and biotic factors alike are looked at from the point of view of the extent to which their effect is to stabilize the equilibrium.[34] The carbon cycle, for instance, can be described as being adjusted by the regulating effects of the oceans and the biological cycle. By means of these powerful theoretical tools, ecology had become the authorized science to describe and explain not only the environment of a single organism or of populations or communities but also geographically larger systems, including Earth as a whole. Thus ecological theories seemed to provide the ideal toolbox to manage any sort of environment, just as cybernetics and general systems theory provided a toolbox to understand, manage and perhaps optimize the behaviour of machines and other technical systems.

In this transition from understanding to managing systems, 'system' became an ambiguous term with a scientific as well as a technical dimension. On the one hand 'system' served as a general representational device for describing and explaining nature and technology as self-contained, conservative, cyclical and self-regulatory processes. On the other hand, if nature, like technological systems, operates in a certain way, this leads to a technical notion of functional systems with performance parameters that can be managed, adjusted and optimized. The powerful promise of GST thus involved a shift from theoretical ecology, based on mathematical modelling, to issues of controlling and managing systems that contain living organisms – a shift from ecoscience to ecotechnologies such as 'space biology' or 'cabin ecology'.

Economy of Spaceship Earth

This 'economy of Spaceship Earth' came to underpin the concerns expressed in the Club of Rome report on the 'Limits to Growth'. And as with cabin ecology in particular, the envisioned control by a few parameters of Spaceship Earth and of planet Earth as a total world model implies a form of excess. Travel into outer space, the current conquest of nanospace and the project of managing the blue planet all share the idea that space itself can be used to exert technical control: within the conservative framework of an absolutely limited Malthusian Earth, the notion of 'carrying capacity' equated available surface area with available space. For example, alarmist images of how much standing room is taken up by all the inhabitants of the Earth translated into political calls for population control underpinned by scientific models. The use of space for technical control came into its own when available surface area became divorced from available space with the notion of the 'ecological footprint'. It also serves to send alarmist messages about the land use required to sustain a single citizen of the US or of India. The measure of the ecological footprint signals that we (in the industrialized countries) are living far beyond our means. At the same time, somewhat paradoxically, it also signals that we *can* live far beyond our means: the sum of ecological footprints already exceeds the available surface area on Earth by a factor of 1.4 – and it is simultaneously the worry of limits-to-growth environmentalists and the hope of technoscientific researchers that this factor will become bigger in years to come.

While excess in molecular biology or in nanotechnology involves shaping the world atom by atom or molecule by molecule, ecotechnology produces excess through manipulation and enhancement of the cybernetic world machine. Today, scientific expertise about the limits to growth serves as a starting point and technological challenge for the so-called sustainability sciences and related technological fields which are primarily concerned with the control, discovery and constant renewal of resources. The declaration of the recently founded World Resource Forum is a good example of this kind of agenda:

> Traditional environmental technologies are no longer enough ... We call for a new global strategy for governing the use of natural resources ... By combining efficiency and resource productivity targets with sufficiency norms evolved through participative mechanisms, it should be possible to avoid the traditional type of growth.[35]

This is a conceptualization of limits that already points at its transgression and therefore exhibits a similar ambivalence as the notion of the self-regulating system. The World Resource Forum asserts that the acknowledgement of limits to resources creates possibilities for escaping these limits by means of efficiency in the sense of enhanced systems performance. This kind of efficiency is to result not primarily from conservation and the avoidance of waste but from technological as well as societal innovation.[36] The programme corresponds to a new environmental movement that embraces technological innovation, being conceptualized for instance as an environmental 'politics of possibility', as discussed at the start of Chapter 11. Other approaches are gathered under the catchwords 'green new deal' or 'green economy', the latter, according to the Rio+20 Conference, seeking 'in principle, to unite under a single banner the entire suite of economic politics and modes of economic analyses of relevance to sustainable development.[37] This has been assessed critically, particularly by political scientists but also by ecological economists, who point mainly to the issue of control expressed, for instance, in the development of 'participative mechanisms'. Further mechanisms of social innovation are suggestions to go beyond the existing sociopolitical limits concerning the externalization of ecological costs from the North to the South, from wealthier to marginalized social groups, or to claim for economic innovation that is to develop regulatory tools face to the unleashed market forces, and more fundamental alternatives such as decoupling wealth, well-being and social quality, or to strengthen the notion of maintenance and of sufficiency.[38]

What can be drawn from this reasoning about the political economy of experiments and the philosophical pattern that was scrutinized in this chapter reflecting the complementarity of science and technoscience (and of course of ecoscience and ecotechnology) in a special and general economy referring to Bataille. Are we confined to the conservative biosphere as outlined by Venadsky? Or does the generous gift of the sun rather produce an abundance and concentration of wealth that needs to be released in the form of excess, waste and creative destruction – such that the technological problem of sustainable development consists in controlling how this release takes place: by grandiose geoengineering schemes, by the construction of optimized landscapes and the enhancement of a green ambient intelligence, or by ever more 'sustainable' forms of production and consumption? Do we accommodate ideas of technological possibility within the framework of knowledge production in the special, restricted 'limited' sciences, or do we view technoscientific research as a productive, creative and 'liberating' force for wealth production?

How might we conceptualize the transformation from a limited world of scarcity to an unbounded world of excess? And how might we control the transformation from a special economics of zero-sum games and of supply balancing demand to a general economics of luxurious abundance and abject waste? It is the exploratory aspects of experimentation and the creative dimension of art and engineering that provide an image of boundless technical innovation which suggests that the world itself is constantly renewable and an unlimited source of novelty.

CONCLUSION: EXPERIMENTS IN PRACTICE – THE WORK OF EXPERIMENTS

At first glance, the phrase 'experiments in practice' may appear to invoke connotations of recent widespread debates about the practice of scientific experiments. Most often such experiments are imagined to occur within the four walls of the science lab; however, they frequently attract even greater attention when they take place – whether more or less intentionally – beyond these walls, such as when chemicals, genetically modified organisms or new technologies are released in an uncontrolled environment. There is an ongoing debate in society about the moral and political implications of such experiments, about who benefits and who suffers as a result of them and, more generally, about the reliability or desirability of scientific knowledge generated this way.

Upon closer inspection, however, 'experiments in practice' may provoke questions relating to its semantic ambiguity and, in particular, about the meaning of 'praxis' contained in it: how are experiments 'put into practice' and what does this suggest for the experimental mode itself? It has been suggested here that probing the experimental mode means, in the first instance, identifying a twofold movement that might be described as (de)liberating the experiment. This refers, first, to the liberation of the scientific experiment from the lab: the move from a lab ideal towards a field ideal. In a second turn it refers also to the deliberation and institution (or establishing) of the social contract between science and society. These two movements are interrelated and belong to an all-encompassing culture of experimentation that stretches from the natural and engineering sciences through virtually every domain of society and even into the fine arts.

A further aspect of 'experiments in practice' refers to a 'renovation' of the role of experiments in the philosophy of science. This change has gradually taken place as experiments have come to be granted 'a life of their own' and new domains such as experimentalism and the philosophy of experimentation have seen the light of day. In the present book, this reorientation is taken as a point of departure to begin delving into the history of the philosophy of science in order to unearth a narrative strand that focuses on 'experiments in practice' and thus on a different philosophy of science.

Francis Bacon's philosophy of experimentation represents an important contribution when it comes to *questioning the scientific method* (discussed in Part I of this book). It helps us understand the culture of experimentation, its tools of inscription and the heuristic methods which not only shape scientific methodology but also deeply pervade all philosophical and social learning. Its linguistic form is the aphorism, which serves – just as the inductive tables do – constantly to challenge the limits of contemporary knowledge. Experiment and observation only became distinguishable when the knowledge of real things became a knowledge of signs that requires experimental probing in order to afford an idea of causal necessity. This is reflected in the reception of Bacon's philosophy in the eighteenth and nineteenth centuries and, above all, in the reception of Hume's problem of induction. Going one step further, Whewell appears as a decisive transitional figure between Baconianism and contemporary philosophy of science. His antithetical philosophy of science is irreducible to matters of justification and evidence. Whewell follows Bacon in placing emphasis on the predominance of practical compared to intellectual skills, which is underwritten by an appreciation for natural history and experimentation in the field.

Different modes of experimentation (discussed in Part II) were developed over time, an observation also reflected in recent philosophy of science. From about the 1980s onwards a new research field known either as experimentalism or the philosophy of scientific experimentation has come into being. Three issues related to this are given particular attention in this part of the book, namely, the role of theory in experimentation, the material realization of the experiment and the complementarity of science and technology. Following this discussion, a scheme is offered that extends the field of experimentation from a lab ideal to a field ideal. The historical co-production of lab and field in the nineteenth century is investigated: as the field sciences became more prominent, the difference to the laboratory was thrown into sharp relief. Snapshots are taken of the experimental settings in physicists' and biologists' laboratories, while the lab bench is considered as a site for experimentation just as the lab as a whole is too. The lab ideal is discussed in detail with reference to the HPS and STS perspective and to varying degrees of specialization. Particular attention is devoted to exploring the field ideal and discussing the requirements of field experiments and research in the field. Field sciences search for and find their objects 'outside', in an uncontrolled environment. Nevertheless, they also perform experiments, and these experiments are necessary in that they cannot be replaced by laboratory practice. The mode of strategic observation, the acknowledgement of place and the idiographic method prove to be essential here.

The field sciences are engaged in *tirelessly tinkering with unruly conditions* (discussed in Part III). Some considerations are offered here regarding the ongoing renewal of the Baconian contract, prompting the question of how far it can be stretched. This is explored by taking a closer look at historical and more recent models and theories of social experimentation. When science acts as an agent of societal change, this entails a transfer of experimental practices to the design, monitoring and evaluation of innovation processes. These processes also involve surprises, and so to enable social learning from collective experiments, a set of procedures and mechanisms for transparency, fairness and responsibility is required. *Transgressing the boundaries* is a central motif here as the experimental mode pushes back the limits, borders or boundaries that define objects in the field. The term 'boundaries' has been used to denote the normative bounds of action while 'border' has been used to describe physical limits. This distinction never caught on, however; instead, border/boundary intermingled in a hybrid-ized discourse that eventually afforded visual metaphors of blurring boundaries and transgressing borders. Both of these have played a role in environmental dis-course over the years, such as in the 'limits to growth' debate and the 'border traumas' of the 1960s and 1970s.

This probing and experimental negotiation of epistemic and disciplinary boundaries is then extended to general accounts of scientific practice in relation to social practice. Two hypotheses structure the discussion here. The first postu-lates that boundaries are coming to the fore because they appear to be vanishing, along with the dualistic pattern that has been taken for granted in modernity. Second, conceptions of border/boundary are rendered in a positive mode in the many accounts of discipline-transgressing research, when creating boundary objects, while doing boundary work and when using boundary concepts in trad-ing or border zones. These transgressive moves can be found in early institutional and practical constellations in ecology, which at the time was known as a 'bridg-ing science' due to basing its epistemic stability explicitly on scientific practice and experimentation even while tinkering with unruly conditions. It is inter-esting to note that since then attention has turned from literally bridging the border/boundaries towards actually constituting the border/boundaries them-selves, a move that has been accompanied by a certain dissolution of disciplines.

Practising experiments in a world of environmental concerns (discussed in Part IV) is conceived as an important tool for tackling environmental problems such as climate change or limitations of resources and space. Accordingly, environ-mental research is one of the innovative forces in modern societies. This brings with it particular experimental practices that are acquiring increasing signifi-cance in the broad field of ecological knowledge production. This is discussed in the context of a conceptual framework that views the vast field of ecological knowledge through the lens of a juxtaposition of two complementary umbrella terms, ecotechnology and ecoscience. The latter is characterized by the develop-ment of general concepts and theories (something that is done in theoretical ecology, for instance) whereas ecotechnology is understood as a technoscience that principally develops local theories and practices.

It can be noted that space (and place) as an oikos in the mode of experimentation is the recurrent theme that links recent debates on green culture issues. This work of conceptualizing ecotechnology through a better understanding of ecotechnological space is pursued by moving through different domains such as restoration ecology, green nanotechnology and cabin ecology.

It is proposed that experimentation may essentially be considered as an economic activity. This activity is not neutral insofar as different modes of housekeeping are put into practice: one is an economy that follows a logic of adaptation to limits while the other is an economy that strives for conquest and thus seeks to transgress limits. It turns out that the exploratory aspects of experimentation ultimately provide an image of boundless technical innovation, one that suggests the world itself is constantly renewable and an unlimited source of novelty.

At first sight it would appear that 'green' technologies are resistant to transgressive experimentalism, mainly because their dominant motivation derives from the limits to growth. It turns out, however, that green technologies are developing strategies to escape these limits and that a moral economy of hedonism can be identified here. A regime of pleasure along with a denial of limits turns out to be characteristic of the technosciences – including the ecotechnologies. The fixtures and fittings of the ecotechnological world are constitutively ambiguous ones and need to be adjusted again and again, thus providing a taste of the likewise pleasurable and precarious work to be done in the context of a knowledge society contract – that is to be administered and cultivated in an experimental mode.

NOTES

Introduction: Towards an Experimental Mode in Science, Society and Philosophy

1. The elusive kernel of a research programme may be encased within a specific natural philosophy: recent scientific concepts such as organism, atom and element are loaded with historical meanings that can be more readily accessed by looking closely at their origins in natural philosophy. This applies equally to the concept of nature in science generally, which is value-laden and implies a subtle normativity. This latter can be revealed and properly analysed only by going back to the moment at which normativity entered the discourse and subsequently came to determine the philosophical debate.

2. These different modes of experimentation will be elaborated in detail in Chapter 7, 'Stretching the Baconian Contract – But How Far?'

3. H. Arendt, 'Social Science Techniques and the Study of Concentration Camps', *Jewish Social Studies*, 12 (1950), pp. 49–64, on p. 60.

4. G. Anders, *Die Antiquiertheit des Menschen: Über die Seele im Zeitalter der zweiten industriellen Revolution* (Munich: Beck, 1956).

5. G. Mitman, *The State of Nature: Ecology, Community, and American Social Thought, 1900–1950* (Chicago: University of Chicago Press, 1992); F. Fischer and M. A. Hajer (eds), *Living with Nature: Environmental Politics as Cultural Discourse* (New York: Oxford University Press, 1999); P. Mcnaghten and J. Urry, *Contested Natures* (London: Sage, 1998); A. Jamison, *The Making of Green Knowledge* (Cambridge: Cambridge University Press, 2001).

6. L. Trepl, *Geschichte der Ökologie: Vom 17. Jahrhundert bis zur Gegenwart* (Frankfurt am Main: Athenäum, 1987).

7. These are just a few exemplary references, somewhat limited to the European context (it is only there that mode 2 research became a relatively influential concept); Gernot Böhme, together with his fellows of the so-called Starnberger MPI group, referred explicitly to the issues put forward by the environmental movement which is fairly visible in his philosophical work (Nowotny and H. Rose (eds), *Counter-Movements in the Sciences: The Sociology of the Alternatives to Big Science* (Dordrecht: D. Reidel, 1979), particularly in that same volume G. Böhme, 'Alternatives in Science – Alternatives to Science?', pp. 105–25; G. Böhme, *Technik, Gesellschaft, Natur: Die zweite Dekade des interdisziplinären Kolloquiums der Technischen Hochschule Darmstadt* (Darmstadt: Technische Hochschule Darmstadt, 1992)). A critical review of some of the main concepts developed during this period is provided by Aant Elzinga (A. Elzinga, 'The New

Production of Particularism in Models Relating to Research Policy: A Critique of Mode 2 and Triple Helix', *Contribution to the 4S-EASST Conference* (Paris, 2004), at www. csi.ensmp.fr/WebCSI/4S/download_paper/download_paper.php?paper= elzinga.pdf [accessed 22 December 2013]).

8.   W. Krohn and W. van den Daele, 'Science as an Agent of Change: Finalization and Experimental Implementation', *Social Science Information*, 37 (1998), pp. 191–222.

9.   G. Böhme and N. Stehr (eds), *The Knowledge Society: The Growing Impact of Scientific Knowledge on Social Relations* (Dordrecht: D. Reidel, 1986); P. Weingart, W. Krohn and M. Carrier (eds), *Nachrichten aus der Wissensgesellschaft. Analysen zur Veränderung der Wissenschaft* (Weilerswist: Velbrück Wissenschaft, 2007).

10. These arguments were developed in a paper by A. Schwarz and W. Krohn, 'Experimenting with the Concept of Experiment: Probing the Epochal Break', in A. Nordmann, H. Radder and G. Schiemann (eds), *Science Transformed? Debating Claims of an Epochal Break* (Pittsburgh, PA: University of Pittsburgh Press, 2011), pp. 119–34.

11. D. T. Campbell, 'Reforms as Experiments', *American Psychologist*, 24 (1969), pp. 409–29, on p. 409.

12. It was Donald Campbell (in ibid.) who set the tone in the rapidly growing field of evaluation research. The field attracted scientists from many different disciplines when experimental approach in/towards society was developed further methodologically and programmatically (H. W. Riecken, *Social Experimentation: A Method for Planning and Evaluating Social Intervention* (New York: Academic Press, 1974); J. Hausman and D. A. Wise, *Social Experimentation* (Chicago,IL: University of Chicago Press, 1985)).

13. S. I. Donaldson, 'In Search of the Blueprint for an Evidence-based Global Society', in S. I. Donaldson, C. A. Christie and M. M. Mark (eds), *What Counts as Credible Evidence in Applied Research and Contemporary Evaluation Practice?* (Newbury Park, CA: Sage, 2008), pp. 2–18, on pp. 2–3.

14. U. Beck, *Risikogesellschaft: Auf dem Weg in eine andere Moderne* (Frankfurt am Main: Suhrkamp, 1986).

15. Accordingly, a German paper published in 1989 and entitled *Gesellschaft als Labor* (*Society as a laboratory*) (W. Krohn and J. Weyer, 'Gesellschaft als Labor: Die Erzeugung sozialer Risiken durch experimentelle Forschung', *Soziale Welt*, 40 (1989), pp. 349–73, English version published in *Science and Public Policy* in 1994) caused no ethical unease in relation to the guinea pig scenario but was showered with praise instead: it became a winner of the 1989 Fritz Thyssen Foundation Award.

16. Ibid., p. 173; B. Wynne, *Risk and Reflexivity: Towards the Retrieval of Human Universals* (London: Sage Publications, 2008).

17. B. Wynne, 'Risk and Environment as Legitimatory Discourses of Technology: Reflexivity Inisde Out?', *Current Sociology*, 50 (2002), pp. 459–76.

18. M. Gross and W. Krohn, 'Science in a Real World Context: Constructing Knowledge through Recursive Learning', *Philosophy Today*, 48 (2004), pp. 38–50; M. Gross, H. Hoffmann-Riem and W. Krohn, *Realexperimente: Ökologische Gestaltungsprozesse in der Wissensgesellschaft* (Bielefeld: Transcript, 2005).

19. U. Beck, 'Die blaue Blume der Moderne', *Der Spiegel*, 33 (1991), pp. 50–1; U. Beck and C. Lau, *Entgrenzung und Entscheidung* (Frankfurt am Main: Suhrkamp, 2004); U. Felt (rapporteur), B. Wynne (chairman), M. Callon, M. E. Gonçalves, S. Jasanoff, M. Jepsen, B. Joly, Z. Konopasek, S. May, C. Neubauer, A. Rip, K. Siune, A. Stirling and M. Tallacchini, *Taking European Knowledge Society Seriously: Report of the Expert Group on Science and Governance to the Science, Economy and Society Directorate* (Brussels: Directorate-

General for Research, European Commission, 2007); J. Habermas, *Moral Consciousness and Communicative Action*, trans C. Lenhardt and S. W. Nicholsen (Cambridge: Polity Press, 1990); J. Habermas, *Der gespaltene Westen: Kleine politische Schriften* (Frankfurt am Main: Suhrkamp, 2004); S. Jasanoff, *Designs on Nature: Science and Democracy in Europe and United States* (Princeton, NJ: Princeton University Press, 2005); A Light, 'Restoring Ecological Citizenship', in B. A. Minteer and B. P. Taylor (eds), *Democracy and the Claims of Nature* (New York: Rowman and Littlefield, 2002), pp. 153–72; A. Nordmann (rapporteur), *Converging Technologies: Shaping the Future of European Societies* (Brussels: EC Report, 2004), at ec.europa.eu/research/conferences/2004/ntw/pdf/final_report_en.pdf [accessed 18 September 2013]; Y. Stavrakakis, 'Passions of Identification: Discourse, Identification, and European Identity', in D. Howarth and J. Torfing (eds), *Discourse Theory in European Politics: Identity, Policy and Governance* (New York: Palgrave, 2005), pp. 68–92; Wynne, 'Risk and Environment as Legitimatory Discourses of Technology'; Wynne, *Risk and Reflexivity*.

20. The snapshot is a type of narrative complementary to the 'big picture' that we expect conventionally to be a description of events over a rather long historical period and with a larger spatial coverage. However, big-picture narratives are not only a question of a quantitative but also of a qualitative difference, as they 'describe things possessed of a certain level of significance, a significance we can say is somewhere on a scale between footwear and the transcendent' (J. R. R. Christie, 'Aurora, Nemesis and Clio', *British Journal for the History of Science*, 26 (1993), pp. 391–405, on p. 392).

21. Felt et al., *Taking European Knowledge Society Seriously*, p. 68; A. Nordmann, 'European Experiments', *Osiris*, 24 (2009), pp. 278–302, on pp. 282–3.

22. C. Waterton and B. Wynne, 'Building the European Union: Science and the Cultural Dimensions of Environmental Policy', *Journal of European Public Policy*, 3 (1996), pp. 421–40.

23. This term was used by Robert E. Kohler in his book *Landscapes and Labscapes: Exploring the Lab-field Frontier in Biology* (Chicago, IL: University of Chicago Press, 2002).

24. These include learned societies and nature conservation organizations.

25. The terms 'ecoscience' and 'ecotechnology' are commonly used, however, in a rather diffuse way. In this book 'ecoscience' is oriented towards a science ideal that is theory or concept oriented, where experiments confirm or refute a hypothesis. 'Ecotechnology' is the complementary technoscientific ideal that is oriented more towards objects and agencies; the function of an ecosystem is claimed to be a possible capacity that may serve as an ecosystem service.

1 (De)Liberating the Experiment: A Different History of the Philosophy of Science

1. G. C. Lichtenberg, *Schriften und Briefe*, ed. W. Promies, 6 vols (Frankfurt am Main: Zweitausendeins, 1994), vol. 1, p. 830 [J 1242]. The word 'hoist' does greater justice to the historical context; however, 'tool' is equally adequate and is thus used in the following. Thanks to Steven Tester for his support in coping with the translation of this aphorism.

2. I. Hacking, *Representing and Intervening* (Cambridge: Cambridge University Press, 1983), p. 150.

3. M. Muntersbjorn, 'Francis Bacon's Philosophy of Science: Machina intellectus and forma indita', *Philosophy of Science*, 70 (2003), pp. 1137–48, on p. 1141; P. Urbach, *Francis Bacon's Philosophy of Science: An Account and a Reappraisal* (Chicago, IL: Open Court, 1987).

4. P. Forman, 'The Primacy of Science in Modernity, of Technology in Postmodernity, and of Ideology in the History of Technology', *History and Technology*, 23 (2007), pp. 1–152, on p. 3. To substantiate his argument, Forman repeatedly points to Lewis Mumford's address delivered at the University of Pennsylvania in Philadelphia on 24 January 1961, to the Conference on the Influence of Science upon Modern Culture, to commemorate the 400th anniversary of the birth of Francis Bacon. The paper was published the same year as 'Science as Technology', *Proceedings of the American Philosophical Society*, 105 (1961), pp. 506–11.

5. In the mainstream history of philosophy of science in the twentieth century, this aspect of Whewell's philosophy was – and still is – largely overlooked in favour of his logical and linguistic discussion of a philosophy of induction.

6. Muntersbjorn, 'Francis Bacon's Philosophy of Science', p. 1139.

7. F. Bacon, *The Instauratio Magna Part II: Novum Organum and Associated Texts*, ed. and trans G. Rees and M. Wakely, 15 vols (Oxford: Clarendon Press, 2004), vol. 11, p. 8. If not otherwise marked, all Bacon quotes refer to this edition, volume 11. The abbreviation used is 'B 11, p. 8', possibly followed by 'a x', giving the number of the aphorism; a2 is used for the second book of aphorisms.

8. In his introduction of *Philosophies of Technology*, Zittel points to the methodological aspects (C. Zittel, 'Introduction', in C. Zittel, G. Engel, R. Nanni and N. C. Karafyllis (eds), *Philosophies of Technology: Francis Bacon and his Contemporaries* (Leiden and Boston, MA: Brill, 2008), pp. 19–29). Bacon's influence on literature is pointed out by L. Jardine, *Francis Bacon: Discovery and the Art of Discourse* (Cambridge: Cambridge University Press, 1974); mainly by C. Schildknecht, 'Experiments with Metaphors: On the Connection between Scientific Method and Literary Form in Francis Bacon', in Z. Radman (ed.), *From a Metaphorical Point of View: A Multidisciplinary Approach to the Cognitive Content of Metaphor* (Berlin and New York: De Gruyter, 1995), pp. 27–50; and also by Muntersbjorn, 'Francis Bacon's Philosophy of Science'.

9. T. Golan, *Laws of Men and Laws of Nature: The History of Scientific Expert Testimony in England and America* (Cambridge, MA: Harvard University Press, 2004).

10. B. J. Shapiro, *A Culture of Fact: England, 1550–1720* (Ithaca, NY: Cornell University Press, 2000).

11. A. Ede and L. B. Cormack, *A History of Science in Society: From Philosophy to Utility* (Peterborough: Broadview Press, 2004), pp. 144–5.

12. The aspect of abusive domination over nature was particularly criticized from a feminist position. Elizabeth Hanson treats torture as a 'paradigm for discovery' (*Discovering the Subject in Renaissance England* (Cambridge: Cambridge University Press, 1998), pp. 20–5); Carolyn Merchant points out that the transition from private secrets to public knowledge in Bacon's *Great Renewal* 'cannot be understood apart from its context of gendered rhetoric' (C. Merchant, 'Secrets of Nature: The Bacon Debates Revisited', *History of Ideas*, 69 (2008), pp. 147–62, on p. 155). Pesic counters that Bacon's reference to the Protean myth is not about the torture of nature but rather a depiction of a heroic mutual struggle, in which the 'nobility both of the seeker and of nature' is tested (P. Pesic, 'Wrestling with Proetus: Francis Bacon and the "Torture" of Nature', *Isis*, 90 (1999), pp. 81–94, on p. 82).

13. M. Poovey, *A History of the Modern Fact* (Chicago, IL: University of Chicago Press, 1998), p. xvii.

14. Ede and Cormack, *A History of Science in Society*, p. 91.

15. F. Bacon, *Neues Organon: Vorrede: Erstes Buch*, ed. W. Krohn, 2 vols (Hamburg: Meiner, 1990), vol. 1, p. xxxi.

16. B 11, p. 65, a 3. In a translation dating from 1676 it says, 'Things are performed by instruments and aids'. One might argue that this shifts attention away from the process to the item afforded (F. Bacon, *The Novum Organum*, trans. M. D. B. D. (London: Thomas Lee at the Turkshead, 1676).

17. B 11, p. 65, a 2.

18. Ibid., p. 205, a2 4.

19. Krohn continues, 'and it remains to be clarified what form the inner coherence takes between interaction with nature and interpretation of nature' (W. Krohn, 'Einleitung', in W. Krohn (ed.), *Francis Bacon: Neues Organon: lateinisch-deutsch*, 2 vols (Hamburg: Meiner, 1990), vol. 1, pp. ix–lvi, on p. xvi; trans. AS). This accords quite well with the conceptual distinction between invention and finding, presented in the following.

20. D. Baird, 'Thing Knowledge', in H. Radder (ed.), *The Philosophy of Scientific Experimentation* (Pittsburgh, PA: University of Pittsburgh Press, 2003), pp. 39–67; B. Bensaude-Vincent, S. Loeve, A. Nordmann and A. Schwarz, 'Matters of Interest: The Objects of Research in Science and Technoscience', *Journal for General Philosophy of Science*, 42 (2011), pp. 365–83.

21. F. Bacon, *The Advancement of Learning*, ed. M. Kiernan, 15 vols (Oxford: Clarendon Press, 2000), vol. 4, p. vii.

22. Ibid., p. 61.

23. The parts given in the Distributio Operis are: the first: *The Partitions of the Sciences*; the second: *Novum Organum, or Directions Concerning the Interpretation of Nature*; the third: *The Phenomena of the Universe, or Natural and Experimental History for the Building Up of Philosophy*; the fourth: *The Ladder of the Intellect*; the fifth: *Precursors, or Anticipations of the Philosophy to Come*; the sixth: *The Philosophy to Come, or the Active Science*. Parts 1 and 2 were published in 1620, as well as the Preliminaries of Part 3 and the Preliminaries to the *Instauratio magna*.

24. B 11, p. 27 (emphasis added).

25. Jardine, *Francis Bacon*, p. 2.

26. B 11, p. 157, a 98.

27. L. Jardine and M. Silverthorne, 'Introduction', in L. Jardine and M. Silverthorne (eds), *Francis Bacon, the New Organon* (Cambridge: Cambridge University Press, 2000), pp. vii–xxviii, on p. xiv.

28. To be sure, with his *bee model* (or at this point rather the three-animal-model), Bacon also distances himself explicitly from the traditional philosophical positions of *empirics* and *rationalists*, or perhaps today's realists and constructivists: 'Those who have dealt with the sciences have either been empirics or dogmatists. The empirics, in the manner of the ant, only store up and use things; the rationalists, in the manner of spiders, spin webs from their own entrails; but the bee takes the middle path: it collects its material from the flowers of field and garden, but its special gift is to convert and digest it' (B 11, p. 153, a 95).

29. Ibid.

30. A. Nordmann, 'Philosophy of Science', in B. Clarke and M. Rossini (eds), *The Routledge Companion to Literature and Science* (London: Routledge, 2010), pp. 362–73, on p. 368.

31. W. Whewell, *Novum Organon Renovatum, being the Second Part of the Philosophy of the Inductive Sciences*, 3rd edn (London: Parker and Son, 1858), p. 44. In the following the abbreviation *NOR* is used for W. Whewell, *Novum Organon Renovatum*.

32. Muntersbjorn, 'Francis Bacon's Philosophy of Science', p. 1142; Schildknecht, 'Experiments with Metaphors', pp. 29–30.

33. Following the dominant metaphor of the *Book of Nature* and of the philosopher as the skilled interpreter of this book, Bacon assumed a close similarity of language and the order of nature when he compared the diversity of natural processes to the diversity of written expressions. He expected that the former can be explained by a small number of causes just as the latter can be explained by the combination of a given number of letters. It appears that this idea was taken up in the artificial language movement in seventeenth-century England (R. Lewis, *Language, Mind and Nature: Artificial Languages in England from Bacon to Locke* (Cambridge: Cambridge University Press, 2007)). Interestingly enough, Bacon thought that the deciphering of the book of nature would be manageable in a few years if only it were funded sufficiently (B. Farrington, *Francis Bacon: Philosopher of Industrial Science* (London: Lawrence and Wishart, 1951)). This is clearly a view of the 'naive empiricist' Bacon.

34. F. Bacon, *De Augmentis*, p. 292, quoted in Krohn, 'Einleitung', p. xxxvii.

35. A. Schöne, *Aufklärung aus dem Geist der Experimentalphysik: Lichtenbergsche Konjunktive* (Munich: C. H. Beck, 1982); F. Steinle, 'Experiment', in F. Jäger (ed.), *Enzyklopädie der Neuzeit*, 3 vols (Stuttgart and Weimar: Metzler, 2006), vol. 3, pp. 722–8.

36. R. M. Schuler, 'Francis Bacon and Scientific Poetry', *Transactions of the American Philosophical Society*, 82 (1992), pp. 1–65, on p. 46.

37. Lichtenberg, *Schriften und Briefe*, vol. 1, p. 830 [J 1242].

38. Schöne describes the aphorism 'Einen Finder zu erfinden für alle Dinge' [J 1621] as Lichtenberg's 'kardinale Suchformel' (most important search formula). The note that follows directly in Lichtenberg's Sudelbücher is 'Ein Tubus Heuristicus' [J 1622] (ibid., vol. 2, p. 297).

39. Schöne, *Aufklärung aus dem Geist der Experimentalphysik*, p. 76.

40. It is most likely that Lichtenberg was familiar with the Kantian distinction between 'ostentatious concepts' and 'heuristic concepts', the latter giving mere principles for a possible search without giving any instructions on how to deduce afterwards (H. Schepers, 'Heuristik, heuristisch', *Historisches Wörterbuch der philosophischen Begriffe*, 13 vols (Basel: Schwabe, 1974), vol. 4, pp. 1115–20, on p. 1118). This is clearly not the concept that Lichtenberg is after when he points to the 'finder' as a practical tool.

41. Popper, chapter 2, sections 5–8, in P. A. Schilpp (ed.), *The Philosophy of Karl Popper* (Chicago, IL: Open Court, 1974). Popper himself would have certainly objected to having his hero scientist appear to be more or less a 'naive empiricist'. However, Popper acknowledged Bacon's active scientist (Urbach, *Francis Bacon's Philosophy of Science*) even if it remains unclear how big the divide is between the 'adventure of brave ideas' and the search for the 'middle path'. To be sure, Popper overlooked the precision that Bacon put into the development of intermediate means – his middle path – for the transformation of sense data into experience.

42. Jardine, *Francis Bacon*, p. 177.

43. Ibid.

44. Krohn, 'Einleitung', p. xxxvii.

45. S. Weigel, 'Das Gedankenexperiment: Nagelprobe auf die facultas fingendi in Wissenschaft und Literatur', in T. Macho and A. Wunschel (eds), *Science and Fiction: Über*

Gedankenexperimente in Wissenschaft, Philosophie und Literatur (Frankfurt am Main: Fischer, 2004), pp. 183–208, on p. 185; A. Schwarz, 'Wilde und liederliche Naturen: Stanislaw Lems nanotechnologische Vorbilder', in D. Korczak and A. Lerf (eds), *Zukunftspotentiale der Nanotechnologien: Erwartungen, Anwendungen, Auswirkungen* (Kröning: Asanger, 2007), pp. 103–26, on pp. 105–6.

46. B 11, p. 159, a 101 (emphasis added).

47. Jardine, *Francis Bacon*, p. 15.

48. *Langenscheidts Taschenwörterbuch Lateinisch-Deutsch*, Vol. 1 (Berlin: Langenscheidt, 1963), trans. AS. Almost exchangeable are the meanings given in an 'enlarged' dictionary based on an edition from 1913 (revised in 1959) where it says 'I. assay' and 'II. experience achieved by proof' for the entry 'experiential', and for the entry 'experimentum' 'I. assay, proof' and 'II. experience' (K. E. Georges, *Ausführliches Lateinisch-Deutsches Handwörterbuch*, 2 vols (Basel: Benno Schwabe, 1959), vol. 1, pp. 2579–81.

49. Georges, *Ausführliches Lateinisch-Deutsches Handwörterbuch*. See also again *Langenscheidts Taschenwörterbuch Latein-Deutsch*, p. 206: to essay/try, to probe, to proof; to compete with s.o.; to try one's power on s.o.; to dispute, to go to court; to take a chance, to jeopardise, to attempt; to learn by experience; to go through s.th., to suffer, to sustain s.th. (trans. AS).

50. A. Walde, *Lateinisches etymologisches Wörterbuch* (Heidelberg: Carl Winter's Universitäts-Buchhandlung, 1939), p. 288. According to that dictionary, *experientia* and *experimentum* first appeared in loci classici by Varro and Cicero.

51. The Baconian *empirics* can be divided up into empiricists, sensationists or ideólogues. All of them insist on the importance of experience, even if their theories 'differ essentially in the way in which they interpret experience' (J. C. O'Neal, *The Authority of Experience: Sensationist Theory in the French Enlightenment* (University Park, PA: Pennsylvania State University Press, 1996), p. 229.

52. G. Pomata, 'Observation Rising: Birth of an Epistemic Genre, 1500–1650', in L. Daston and E. Lunbeck (eds), *Histories of Scientific Observation* (Chicago, IL: University of Chicago Press, 2011), pp. 45–80, on p. 45.

53. F. Steinle, *Explorative Experimente: Ampère, Faraday und die Ursprünge der Elektrodynamik* (Wiesbaden: Franz Steiner, 2005), p. 723.

54. L. Jardine and M. Silverthorne (eds), *Francis Bacon, the New Organon* (Cambridge: Cambridge University Press, 2000), p. 57 n. 31.

55. B 11, p. 205, a2 4.

56. Ibid., p. 77, a 34.

57. Ibid., pp. 77–8, a 36.

58. Quote taken from the Oxford University Press online announcement of *The Oxford Francis Bacon* (vol. 12) at http://www.oup.com.

59. Krohn, 'Einleitung', p. xxx.

60. B 11, p. 79, a 40 (emphasis in original).

61. Ibid., p. 205, a2 4.

62. Ibid., p. 255, a2 17.

63. R. Harré, 'Recovering the Experiment', *Philosophy*, 73 (1998), pp. 353–77; R. Harré, 'The Materiality of Instruments in a Metaphysics for Experiments', in H. Radder (ed.), *The Philosophy of Scientific Experimentation* (Pittsburgh, PA: University of Pittsburgh Press, 2003), pp. 19–38.

64. G. Bachelard, *L'activité rationaliste de la physique contemporaine* (Paris: Presses universitaires de France, 1951).

65. The term affordance was introduced by James Gibson (*The Ecological Approach to Visual Perception* (Boston, MA: Houghton Mifflin, 1979)), who proposed a focus on the agency of an object and its environment. It was also adopted to distinguish two modes of knowledge, the scientific and the technoscientific, and above all the role of objects in these ontologically and epistemologically quite different concepts of relatedness (Bensaude-Vincent et al., 'Matters of Interest').

66. B 11, p. 81, a 41.

67. Ibid., p. 87, a 50.

68. Ibid., p. 33.

69. Ibid., p. 87, a 50.

70. Ibid., a 49.

71. Just as a reminder of his worry about the confusion and unreliability of a data inventory: 'Only when that becomes standard practise, with experience at last becoming literate, should we hope for better things' (ibid., p. 159, a 101).

72. Ibid., p. 253, a2 15.

73. Ibid., a2 16.

74. Ibid., p. 137, a2 14.

75. Ibid., p. 253, a2 19.

76. *NOR*, p. 108.

77. B 11, p. 71, a 19.

78. Ibid., p. 161, a 104.

79. Ibid., p. 163, a 104.

80. P. Smith, *The Body of the Artisan: Art and Experience in the Scientific Revolution* (Chicago, IL: Chicago University Press, 2004), p. 233.

81. Ibid.; see also L. Daston and E. Lunbeck (eds), *Histories of Scientific Observation* (Chicago, IL: University of Chicago Press, 2011).

82. Gibson, *The Ecological Approach to Visual Perception*, p. 129.

83. Ibid., p. 240. Gibson explicitly creates a link to ecology, which was at the time the most important discipline (and also ideology) that embodied this kind of relational thinking: '[Affordances] are ecological, in the sense that they are properties of the environment relative to an animal' (J. J. Gibson, 'New Reasons for Realism', *Synthese*, 17 (1967), pp. 401–18, on p. 404).

84. L. Daston, 'The Empire of Observation, 1600–1800', in L. Daston and E. Lunbeck (eds), *Histories of Scientific Observation* (Chicago, IL: University of Chicago Press, 2011), p. 86.

85. B 11, p. 65, a 4.

2 The Philosophy of Inductive Sciences

1. The first part, *History of the Inductive Sciences*, appeared in 1837 and *Philosophy of the Inductive Sciences, Founded upon their History* followed in 1840; this was enlarged and divided into three parts and published as the *History of Scientific Ideas* (2 vols; 1858), *Novum Organum Renovatum* (1858) and *The Philosophy of Discovery* (1858).

2. It has been noted by a number of authors that Whewell comes closer to Kant than he himself ever made explicit.

3. W. Whewell, *Aphorisms Concerning Ideas, Science, and the Language of Science* (La Vergne, TN: Kessinger Publishing, 1840), p. 2, a VIII (emphasis in original).

4. M. Fisch and S. Schaffer, *William Whewell: A Composite Portrait* (Oxford: Clarendon Press, 1991), p. 65.
5. Whewell, *Aphorisms*, p. 1, a IV (emphasis in original).
6. A. W. Heathcote, 'William Whewell's Philosophy of Science', *British Journal for the Philosophy of Science*, 4 (1954), pp. 302–14, on p. 303.
7. W. Whewell, *Philosophy of the Inductive Sciences: Founded upon their History*, 2 vols, 2nd edn (London: Parker and Son, 1847), vol. 1, p. 24.
8. Ibid., p. 184, in Section 10 'The Fundamental Antithesis Inseparable' (emphasis in original).
9. W. Whewell, *History of the Inductive Sciences*, 2 parts, 3rd edn (London: Parker and Son, 1857; repr. London: Frank Cass & Co. Ltd, 1967), part 1, pp. 7, 8.
10. Quoted in L. J. Snyder, *Reforming Philosophy: A Victorian Debate on Science and Society* (Chicago, IL: University of Chicago Press, 2006), p. 147. From 1828 to 1832 Whewell held the chair of mineralogy at Cambridge. During this time he produced an essay on mineralogical classification and experiments in a Cornish mine (together with G. B. Airy). Its purpose was to determine the density of the Earth. In June 1838 he was elected to the Knightbridge Chair of Moral Philosophy, which he held until 1855.
11. N. Jardine, J. A. Secord and E. C. Spary, *Cultures of Natural History* (Cambridge: Cambridge University Press, 1996); D. L. Livingstone, *Putting Science in its Place: Geographies of Scientific Knowledge* (Chicago, IL: University of Chicago Press, 2003); U. Klein, *Experiments, Models, Paper Tools* (Stanford, CA: Stanford University Press, 2003); see also Chapter 5 below on 'Experimentation in Lab and Field'.
12. Pomata, 'Observation Rising', p. 50.
13. Ibid., p. 47.
14. *NOR*, p. 59.
15. Daston, 'The Empire of Observation, 1600–1800', p. 87.
16. Translated by Kathleen Cross; 'Versuch, heisset die Erfahrung, so man von einer Sache bekommt, indem solche durch unsern Fleiß hervorgebracht wird. Es ist der Observation entgegen gesetzt, die eine Erfahrung ist, welche uns die Natur freiwillig an die Hand giebet ... allein dergleichen Observationes sind nicht allezeit zulänglich, die wahre Beschaffenheit zu entdecken ... Man muss dahero sich Mühe geben zu versuchen, ob man die Natur nicht dahin vermögen könne, daß sie uns dasjenige sehen lasse, was zu unserm Unterrichte dienet. Dieses erhält man nun durch die *Experimente*', and he continues 'da man durch verschiedene Machinationes und Combination verschiedener Dinge bemercket, was sich vor ein Unscheid in einem Effecter ereignet, wenn diese oder jene Sache zugegen oder abwesend ist; woraus man endlich erlernet, was bey einer Würckung wesentlich oder nicht sey, wodurch man endlich zu der wahren Ursache, die den Effect hervorgebracht hat, selbst gelanget' (J. H. Zedler, *Grosses vollständiges Universal-Lexicon aller Wissenschaften und Künste welche bisshero durch menschlichen Verstand und Witz erfunden und verbessert worden*, 64 vols (Halle and Leipzig, 1734), vol. 8, pp. 2344–5). There is also an entry for 'experiential' but the reader is directed from there to *Erfahrung* (experience).
17. Interestingly enough, the German word *Beobachtung* is neither an entry nor does it appear in the article.
18. Translated by Kathleen Cross; 'welche von den Experimenten oder Versuchen unterschieden, und solchen an die Seite gesetzt ist ... [und auch] gemeine Erfahrung genennet wird, (ist) nichts anders, als eine durch Hülffe der Sinne erkannte Wahrheit von dem, was sich in einzelnen Dingen befindet und zuträget, ohne daß solches allererst durch angewandten Fleiß zum Vorschein oder zur Würcklichkeit gebracht worden. Oder

kürzer: Eine Observation ist eine bloß durch Hülffe der Sinnen bemerckte Eigenschafft oder Begebenheit in einem einzelnen Dinge' (Zedler, *Grosses vollständiges Universal-Lexicon*, vol. 25, p. 278).

19. In the reception of Bacon, the different patterns of perception depended largely on the philosophical background and mainstream approach that was dominant in the respective cultural, linguistic or national context (Jardine and Silverthorne, 'Introduction'; C. Zittel, G. Engel, R. Nanni and N. C. Karafyllis (eds), *Philosophies of Technology: Francis Bacon and his Contemporaries* (Leiden and Boston, MA: Brill, 2008).

20. G. Tweedale, '"Days at the Factories": A Tour of Victorian Industry with the Penny Magazine', *Technology and Culture*, 29 (1988), pp. 888–903, on p. 889.

21. Ibid., p. 896.

22. Society for the Diffusion of Useful Knowledge, 'An Account of Lord Bacon's *Novum organon scientarium* or, New Method of Studying the Sciences', in C. Knight (ed.), *Library of Useful Knowledge* (London: Baldwin and Cradock, 1828), pp. 1–40.

23. Ibid., p. 3. The editor Charles Knight published the *Library of Useful Knowledge* under the superintendence of the Society for the Diffusion of Knowledge (since 1832, it was known as the *Penny cyclepaedia of the Society for the Diffusion of Useful Knowledge*).

24. Ibid., p. 940.

25. Ibid. p. 368.

26. B. Russell, *History of Western Philosophy and its Connection with Political and Social Circumstances from the Earliest Times to the Present* (London: Allen and Unwin, 1946), pp. 685, 698.

27. M. Diderot and M. d'Alembert (eds), *Encyclopédie, ou, Dictionnaire raisonné des sciences, des arts et des métiers* (Lausanne and Berne: Sociétés typographiques, 1779–82), vol. 2, p. 10 (electronic version).

28. Ibid., p. 9.

29. D. Hume, *A Treatise of Human Nature: Being an Attempt to Introduce the Experimental Method of Reasoning into Moral Subjects*, 3 vols (London: Printed for John Noon, 1739–40), vol. 1, pp. 6–7.

30. T. Cathcart and D. Klein, *Plato and a Platypus Walk into a Bar. Understanding Philosophy through Jokes* (New York: Penguin Books, 2007), p. 57.

31. See for instance A. F. Chalmers, *What is this Thing Called Science?* (1976; Queensland: University of Queensland Press, 1999), p. 45, who provides no reference.

32. Hume himself never denoted the problem as such. Also, he used the term induction very rarely and not in the sense of a logical operation: 'It requires scarce any induction to conclude from hence, that the *idea*, which we form of any finite quality, is not infinitely divisible, but that by proper distinctions and separations we may run up this idea to inferior ones, which will be perfectly simple and indivisible. In rejecting the infinite capacity of the mind, we suppose it may arrive at an end in the division of its ideas; nor are there any possible means of evading the evidence of this conclusion' (Hume, *A Treatise of Human Nature*, pp. 54–5; emphasis in original).

33. B. Russell, *The Problems of Philosophy* (Oxford: Oxford University Press, 1997), p. 63.

34. Hume, *A Treatise of Human Nature*, p. 236.

35. Russell, *The Problems of Philosophy*, p. 63.

36. For a more extended discussion of this argument, see P. Prevos, *Origins of Modern Philosophy: The Problem of Induction* (May 2005), at http://www.prevos.net [accessed 4 February 2011], pp. 4–5.

37. Russell, *The Problems of Philosophy*, pp. 68, 69.

38. Russell, *History of Western Philosophy*, p. 689.
39. In general, demonstrative knowledge is opposed to knowledge based on persuasion, or even probable proof. In a demonstration (or proof), a connection (or relation) between the premise and the conclusion is secured by other premises and by the laws of logic. Aristotle described a demonstration as a step-by-step deduction whose premises are known to be true. The scholastic conception claimed that knowledge is a demonstration of or from first principles. With his concept of knowledge Hume remained committed to this. Locke qualified demonstrative knowledge as being less certain than intuitive knowledge, because it requires effort and attention to go through the steps needed to recognize the connections.
40. Hume, *A Treatise of Human Nature*, p. 7.
41. Ibid., p. 159 (emphasis in original).
42. Ibid., p. 160.
43. Russell, *History of Western Philosophy*, p. 689.
44. Hume, *A Treatise of Human Nature*, p. 274 (emphasis added).
45. I. Hacking, 'Another World is being Constructed Right Now: The Ultracold', in H.-J. Rheinberger (ed.), *Conference 'the Shape of Experiment'*, Berlin, 2–5 June 2005, Preprint 318 (Berlin: MPI für Wissenschaftsgeschichte, 2006), pp. 15–44, on p. 18. Hacking elaborates at some length how precisely Hume refines the scholastic conceptions of opinion versus knowledge (as demonstration from first principles) by stripping knowledge of causation. 'In the writing of Hume, the term "knowledge" is reserved for pure mathematics' (Hacking, *Representing and Intervening*, p. 180).
46. M. Foucault, *Les mots et les choses* (Paris: Gallimard, 1966), p. 73.
47. G. Berkeley, *A Treatise Concerning the Principles of Human Knowledge – Part I: Wherein the Chief Causes of Error and Difficulty in the Sciences, with the Grounds of Scepticism, Atheism, and Irreligion, are Inquir'd* (Dublin: Jeremy Pepyat, 1710), reproduction of original from British Library, University of Basel online, at http://find.galegroup.com/ ecco/infomark.do?&source=gale&prodId=ECCO&userGroupName=unibas&tabID =T001&docId=CW118263402&type=multipage&contentSet&ECCOArticles&ver sion=1.0&docLevel=FAACSIMILE [accessed 3 March 2011], sec. 65.
48. Hacking, *Representing and Intervening*, p. 183.
49. Foucault, *Les mots et les choses*, p. 74; English translation from M. Foucault, *The Order of Things: An Archaeology of the Human Sciences* (New York: Pantheon Books, 1970), p. 60.
50. Hacking, *Representing and Intervening*, p. 183.

3 Whewell's Innovation through Renewal

1. G. Steiner, *Dix raisons (possibles) à la tristesse de pensée*, ed. and trans. P.-E. Dauzat (Paris: Albin Michel, 2005), p. 72. In French it says: 'Où elle vise, où elle invoque la "vérité", la pensée relativise ce critère à l'instant même où elle s'en réclame. Il n'y a pas moyen d'échapper à ce cercle dialectique' (ibid., p. 73).
2. Snyder, *Reforming Philosophy*, p. 21.
3. A recent remark by William Wimsatt reads in a strikingly similar way. In his book *Re-engineering Philosophy for Limited Beings* (Cambridge, MA: Harvard University Press, 2007), he argues that philosophers of science should ideally take their problems from science (and nature) and not so much from analytical philosophy; see also J. R. Griesemer, 'Taking Organic Processes: Representations and Research Styles in Classical Embryol-

ogy and Genetics', in M. D. Laubichler and J. Maienschein (eds), *From Embryology to Eco-devo* (Cambridge, MA: MIT Press, 2007), pp. 375–433, on pp. 380–1.

4.   Snyder, *Reforming Philosophy*, p. 90.

5.   Fisch and Schaffer, *William Whewell*, p. 64.

6.   W. Whewell, *Theory of Scientific Method*, ed. R. E. Butts (Indianapolis, IN: Hackett Publishing, 1989), p. 58.

7.   Whewell chose this metaphor in a review with the programmatic title *Modern Science – Inductive Philosophy*, in which he discussed a book by fellow Breakfast Club member John Herschel, namely the *Preliminary Discourse on the Study of Natural Philosophy* (W. Whewell, 'Preliminary Discourse on the Study of Natural Philosophy', *Quarterly Review*, 45 (1831), pp. 374–407).

8.   Whewell, *History of the Inductive Sciences*, part 1, p. 13.

9.   *NOR*, p. 60.

10.   Whewell, *Aphorisms*, p. 27, a 34.

11.   *NOR*, p. 86 (emphasis in original).

12.   R. E. Butts, 'Introduction', in W. Whewell, *Theory of Scientific Method*, ed. R. E. Butts (Indianapolis, IN: Hackett Publishing, 1989), p. 19; for Whewell, Kepler's law (describing the elliptical orbits of planets) is far beyond a simple description of what is 'there', simply apparent by observation, obtained by a measurement, or merely calculated. Accordingly, 'The Inductive truth is never the mere *sum* of the facts' (*NOR*, p. 108; emphasis in original).

13.   *NOR*, p. 110 (emphasis in original).

14.   Whewell, *Aphorisms*, p. 27, a 37.

15.   Ibid., p. 26, a 30.

16.   *NOR*, p. 164.

17.   Ibid., p. 272. The power of Linnean systematics has been called 'linguistic imperialism' – which might have contributed to Whewell's fascination. Londa Schiebinger has pointed out that the development of this botanical system was part of 'a politics of naming that accompanied and promoted European global expansion and colonization' in the eighteenth century (L. Schiebinger, 'Naming and Knowing: The Global Politics of Eighteenth-Century Botanical Nomenclatures', in P. Smith and B. Schmidt (eds), *Making Knowledge in Early Modern Europe: Practices, Objects, and Texts 1400–1800* (Chicago, IL: Chicago University Press, 2007), pp. 90–105, on p. 91). In her comparative analysis she shows that the Linnean system was so successful not because it was epistemologically required but because for many historical reasons it 'has come into common usage, and its emergence illustrates the importance of understanding knowledge making as a social process' (ibid., p. 105).

18.   *NOR*, p. 147 (emphasis in original).

19.   Ibid.

20.   Ibid., p. 289.

21.   Ibid., p. 147. His concern is justified, of course, by the very visible progress of arithmetic and statistics in the context of nineteenth-century industrialization and global topographical surveys.

22.   Debates on a missing link either in evolutionary biology or geology are examples for this kind of prediction.

23.   *NOR*, p. 21, a XCII.

24.   Ibid., p. 22, a XCIII.

25.   Ibid., a XCIV.

26. Whewell, 'From the Mechanical Euclid, Conataining the Elements of Mechanics and Hydrostatics' (1837), in Whewell, *Theory of Scientific Method*, p. 47.
27. *NOR*, p. 224.
28. Ibid., p. 221.
29. Ibid.
30. Ibid., p. 81.
31. A. Schwarz and A. Nordmann, 'Unlimited Potential in a Limited World: The Political Economy of Eco- and Nanotechnologies', in M. Carrier and A. Nordmann (eds), *Science in the Context of Application: Methodological Change, Conceptual Transformation, Cultural Reorientation* (Berlin: Springer, 2010), pp. 317–36.
32. *NOR*, p. 224.
33. Ibid.
34. Ibid.
35. Ibid.
36. Ibid. (emphasis added).
37. Ibid., p. 225.
38. Ibid., p. 226.
39. Ibid., p. 225.
40. Ibid., p. 114 (emphasis in original).
41. Ibid., p. 108 (emphasis in original).
42. Whewell, *History of the Inductive Sciences*, part 1, p. 14.
43. See, for instance, the following section in *NOR*, p. 192: 'Huyghens, more happy, introduced the idea of the *axis of symmetry* of the solid, and thus was able to give the true law of the phenomena. 8. Although the selected idea is proved to be the right one, only when the true law of nature is established by means of it, yet it often happens that there prevails a settled conviction respecting the relation which must afford the key to the phenomena, before the selection has been confirmed by the laws to which it leads. Even before the empirical laws of the tides were made out, it was not doubtful that these laws depended upon the places and motions of the sun and moon'.
44. Ibid., p. 194.
45. L. Laudan, 'William Whewell on the Consilience of Inductions', *Monist*, 55 (1971), pp. 368–91, on p. 368. Butts notes that Whewell's 'philosophical instincts are all on the side of realism' and that 'it appears unlikely that he would have accepted any form of instrumentalism or antirealism as a correct theory of science' ('Introduction', p. 26). This argument is mainly put forward to counter an overemphasis on 'psychological' features in Whewell's theory of science, such as the identification of laws of nature with laws of thought. Butts suggests instead that it is better to follow Schlick and to interpret apparently psychological matters in a context of logic and epistemology, which is also helpful in linking Whewell's philosophy to canonical philosophy of science.
46. Laudan ('William Whewell on the Consilience of Inductions', p. 370) points out that Whewellian induction is similar to what Peirce later called abduction, that is, the act of being prompted by a surprise to find some general hypothesis which accounts for the known facts. 'Deduction proves that something must be; Induction shows that something actually is operative; Abduction merely suggests that something may be' (C. S. Peirce, *Pragmatism and Pragmaticism: Collected Papers of Charles Sanders Peirce*, 8 vols (Cambridge, MA: Harvard University Press, 1934), vol. 5, p. 171.
47. Hypotheses are, as it were, the vademecum of science, because 'rightly used [they] are among the helps, far more than the dangers, of science; – that scientific induction is not

a "cautious" or a "rigorous" process in the sense of abstaining from such suppositions, but in not adhering to them till they are confirmed by fact, and in carefully seeking from facts confirmation or refutation' (*NOR*, p. 82).

48. To remind the reader of the prevalence of mind over empirical data in Whewell's philosophy, here once again the phrase already quoted above: 'In each inductive process, made by Induction, there is some general idea introduced, which is given not by the phenomena, but by the mind' (ibid., p. 73). Whewell points to the fact that he quotes from his article 'From the Mechanical Euclid' (see n. 26 above).

49. S. Gaukroger, *Francis Bacon and the Transformation of Early-Modern Philosophy* (Cambridge: Cambridge University Press, 2001), pp. 138–9.

50. Ibid., p. 140.

51. S. Jasanoff, *States of Knowledge: The Co-Production of Science and the Social Order* (London: Routledge, 2006), or also S. B. Gill and J. Borchers, 'Knowledge in Co-Action', in S. B. Gill (ed.), *Cognition, Communication, and Interaction* (Dordrecht: Springer, 2008), pp. 38–55.

52. *NOR*, p. 169.

53. Ibid., p. 160.

54. Ibid., p. 244.

55. Today, most historians and philosophers of science would probably object that the steam engine is not a good example for the model that engineering follows upon science.

56. Laudan, 'William Whewell on the Consilience of Inductions', p. 385.

57. 'Consilience of inductions' is usually translated into German simply as 'Übereinstimmung von Induktionen'.

58. Butts, 'Introduction', p. 29.

59. *NOR*, p. 96.

60. Lakatos calls Whewell a 'conservative conventionalist', namely a conventionalist who adheres to the idea that the growth of an empirical science (a 'science of facts') is dependent on the truth of objective facts (*The Methodology of Scientific Research Programmes*, Vol. 1, ed. J. Worral and G. Currie (Cambridge: Cambridge University Press, 1978), p. 106, footnote). According to Lakatos, such a conventionalist must identify a metaphysical principle that ultimately 'superimposes the rules for the game of science'. As an example of such a metaphysical principle, Lakatos mentions principles of induction such as those entailed by Popper's test statements and his corresponding regulative idea of truth. Another example is the idea of truth as being socially constituted: if trained ('purified') scientists agree on certain facts, then those facts are scientifically true (I. Lakatos, 'Falsification and the Methodology of Scientific Research Programmes', in I. Lakatos and A. Musgrave (eds), *Criticism and the Growth of Knowledge* (Cambridge: Cambridge University Press, 1970), pp. 91–195.

61. Snyder, *Reforming Philosophy*, p. 174.

62. Whewell, *History of the Inductive Sciences*, part 1, p. 9.

63. Of course, Lakatos's research programmes offer a much more sophisticated approach to the role of ad-hoc hypotheses as a means of compensating for anomalies at the empirical level and for 'saving' the theoretical construction. However, the dominance of concepts and the 'colligation' over mere perception and the empirical level is similar. As he points out, the hallmark of empirical progress is not trivial verification, and '"refutations" are not the hallmark of empirical failure', instead 'all scientific programmes grow in a permanent ocean of anomalies. What really count are dramatic, unexpected, stunning

predictions: a few of them are enough to tilt the balance' (*The Methodology of Scientific Research Programmes*, p. 6).

64. *NOR*, p. 90.
65. Ibid., p. 88.
66. Laudan, 'William Whewell on the Consilience of Inductions', p. 372.
67. Popper proposes that the severity of a test of a theory can be used to define and measure the theory's power of explanation (K. Popper, 'Das Wachstum der wissenschaftlichen Erkenntnis', in D. Miller (ed.), *Karl Popper Lesebuch* (Tübingen: Mohr, 1995), pp. 154–63, on pp. 159–60).
68. Laudan, 'William Whewell on the Consilience of Inductions', pp. 388–9.
69. In 'borrowing' the problem of consilience and giving it a different semantic turn without being aware of or barely referring to Whewell's work, Popper is not an isolated case in the community of philosophers. Peirce, for example, is very much concerned with the problem of consilience, but only rarely calls it such. In his methodological writings, he reformulates what was identified above as types two and three of the conception of consilience, namely: new predictions based upon the hypothesis are also found to be verified (2) and the hypothesis must be able to explain surprising facts (3).
70. B. M. Stafford, 'Thoughts Not Our Own: Whatever Happened to Selective Attention?', *Theory, Culture and Society*, 26 (2009), pp. 275–93, on p. 285; she refers to the biological theorist Ernst Mayr who had already warned of the attempts of Wilson to reduce biology to physics using consilience as a scheme for an interscientific endeavour.
71. E. O. Wilson, *Consilience: The Unity of Knowledge* (New York: Alfred A. Knopf, 1998), p. 8.
72. Ibid., p. 266.
73. Whewell, *History of the Inductive Sciences*, part 1, p. 11; Whewell refers repetedly to the metaphor of the map of a river, representing the 'Table of the progress of any science ... in which the waters from separate sources unite and make rivulets, which again meet with rivulets from other fountains, and thus go on forming by their junction trunks of a higher and higher order' (ibid.).
74. L. D. Walls, *Consilience Revisited*, electronic book review (December 1999), at www.electronicbookreview.com/thread/criticalecologies/biophilial [accessed 30 October 2013].
75. Ibid.
76. B. M. Stafford, *Visual Analogy: Consciousness as the Art of Connecting* (Cambridge, MA: MIT Press, 2001), p. 105.
77. Ibid., p. 9.
78. The conception of consilience can't be reduced to a mere theory of justification. The fact that Whewell, in his systematic collection of methods, puts some emphasis on the principles of ressemblance (the method of gradation and the method of natural classification) instead of privileging the representational power of numbers and arithmetic gives further evidence.
79. *NOR*, p. 54.
80. Laudan, 'William Whewell on the Consilience of Inductions', p. 391.
81. Ibid.
82. *NOR*, p. 76 (emphasis in original).
83. M. S. Reidy, G. Kroll and E. M. Conway, *Exploration and Science: Social Impact and Interaction* (Santa Barbara, CA: ABC-CLIO, 2007), p. 82.
84. Ibid., p. 85.

85. The observations were made 8–28 June 1835 and were conducted in 101 ports in 7 continental European countries (Spain, Portugal, France, Belgium, Netherlands, Denmark, Norway), 28 in North America, and 573 in the British Isles, including Ireland. In total over 40,000 measurements were collected (Snyder, *Reforming Philosophy*, p.150).

86. W. Whewell, 'Researches on the Tides: Fourth Series: On the Empirical Laws of the Tides in the Port of Liverpool', *Philosophical Transactions of the Royal Society of London*, 126 (1836), pp. 1–15, on p. 13.

87. Ibid.

88. Ibid., p. 2.

89. S. Schaffer, 'The History and Geography of the Intellectual World: Whewell's Politics of Language', in M. Fisch and S. Schaffer (eds), *William Whewell: A Composite Portrait* (Oxford: Clarendon Press, 1991), pp. 201–32, on pp. 218–19.

90. Snyder, *Reforming Philosophy*, p. 251.

91. *NOR*, p. 123; M. Fisch, 'Antithetical Knowledge', in M. Fisch and S. Schaffer (eds), *William Whewell: A Composite Portrait* (Oxford: Clarendon Press, 1991), pp. 289–310, on p. 298; for other scientific concepts, see Snyder, *Reforming Philosophy*.

92. Schaffer, 'The History and Geography of the Intellectual World', p. 201. This was the metaphor used by Trinity fellow John Herschel, who saw the strength of Whewell's inductive philosophy precisely in its linguistic power to make 'scientific language coldly mimetic' (ibid.).

93. Ibid.

94. P. Achinstein, 'The War on Induction: Whewell Takes on Newton and Mill (Norton Takes on Everyone)', *Philosophy of Science*, 77 (2010), pp. 728–39.

95. J. S. Mill, *System der deductiven und inductiven Logik*, 2 vols, trans. J. Schiel (Braunschweig: Vieweg und Sohn, 1868), vol. 1, p. 445.

96. *NOR*, p. 163.

97. Ibid., pp. 80–1.

98. Ibid., p. 160.

99. Ibid.

100. M. Heidelberger, 'Die Erweiterung der Wirklichkeit im Experiment', in M. Heidelberger and F. Steinle (eds), *Experimental Essays: Versuche zum Experiment* (Baden-Baden: Nomos, 1998), pp. 71–92, on p. 89 (trans. AS). This is the last sentence and final conclusion offered by Heidelberger in a paper in which, as it happens, he does not even mention Whewell.

101. *NOR*, pp. 184–5.

102. J. D. Norton, 'A Material Theory of Induction', *Philosophy of Science*, 70 (2003), pp. 647–70, on p. 669. Norton, too, did not mention Whewell at all.

103. Compare P. Galison, *How Experiments End* (Chicago, IL: University of Chicago Press, 1987).

104. H.-J. Rheinberger (ed.), *Conference 'the Shape of Experiment'*, Berlin, 2–5 June 2005, Preprint 318 (Berlin: MPI für Wissenschaftsgeschichte, 2005), p. 3.

4 The Experiment in Recent Philosophy of Science

1. As Peter Janich has complained, a notable exception to this can be seen in the German tradition of constructivism that goes back to Hugo Dingler's book, already published in 1928, *Das Experiment – sein Wesen und seine Geschichte* (*The Experiment – Its Nature and History*) and also includes Janich's own work (P. Janich, 'Was macht experimentelle

Resultate empiriehaltig? Die methodisch-kulturalistische Methode im Experiment', in M. Heidelberger and F. Steinle (eds), *Experimental Essays: Versuche zum Experiment* (Baden-Baden: Nomos, 1998), pp. 93–114). Here, certain experimental procedures are taken to be constitutive of concepts and measurements in the sciences – that is, this view shares with strictly conceptual approaches that it reconstructs the hierarchical architecture of bodies of knowledge. For this reason and the failure of this tradition to influence the mainstream of philosophy of science, I am not considering this here.

2. Hacking, *Representing and Intervening*, p. 150.
3. Hacking, 'Another World is being Constructed Right Now: The Ultracold', p. 16.
4. Hacking, *Representing and Intervening*, p. 149.
5. Ibid., p. 152.
6. Ibid., pp. 154–5.
7. To name just a few edited volumes: H. LeGrand (ed.), *Experimental Inquiries: Historical, Philosophical and Social Studies of Experimentation in Science* (Dordrecht: Kluwer, 1990); A. Pickering (ed.), *Science as Practice and Culture* (Chicago, IL: University of Chicago Press, 1991); M. Heidelberger and F. Steinle (eds), *Experimental Essays: Versuche zum Experiment* (Baden-Baden: Nomos, 1998); M. C. Galavotti, (ed.), *Observation and Experiment in the Natural and Social Sciences* (Dordrecht: Kluwer Academic Publishers, 2003); H. Radder (ed.), *The Philosophy of Scientific Experimentation* (Pittsburgh, PA: University of Pittsburgh Press, 2003).
8. Heidelberger and Steinle (eds), *Experimental Essays*.
9. Radder (ed.), *The Philosophy of Scientific Experimentation*.
10. K. A. Appiah, *Presidential Address*, Eastern Division APA (December 2007), at http://appiah.net/wp-content/uploads/2010/10/APA-Lecture-2007-for-Web.pdf [accessed 12 September 2013]. See also his book on *Experiments in Ethics* (Cambridge, MA: Harvard University Press, 2008).
11. An inventive angle on this has been put forward by Michael Heidelberger. He claims that observation statements are the only point at which experiment and theory enter into any kind of connection with one another. 'Experiment and theory impact on one another only in the few brief encounters that involve verification' (Heidelberger, 'Die Erweiterung der Wirklichkeit im Experiment', p. 72).
12. T. S. Kuhn, *The Essential Tension: Selected Studies in Scientific Tradition and Change* (Chicago, IL: University of Chicago Press, 1977), p. 45.
13. Ibid., p. 48.
14. Ibid., p. 64.
15. H. von Helmholtz, *Vorträge und Reden*, 2 vols (Braunschweig: Vieweg und Sohn, 1903), p. 237. He presented these thoughts in a talk with the programmatic title 'The Facts in Perception' at the Stiftungsfeier der Friedrich-Wilhelms-Universität Berlin (today Humboldt Universität Berlin) in 1878. The manuscript was only published in 1903.
16. E. Zahar, 'Logic of Discovery or Psychology of Invention?', *British Journal for the Philosophy of Science*, 34 (1983), pp. 243–61, on p. 243.
17. R. Hilpinen, 'On Experimental Questions', in D. Batens and J. P. v. Bendegem (eds), *Theory and Experiment: Recent Insights and New Perspectives on their Relation* (Dordrecht: Reidel, 1988), pp. 15–32, on p. 25.
18. H. A. Simon, 'Discovery, Invention, and Development: Human Creative Thinking', *Proc. Nat. Acad. Sci. USA*, 80 (1983), pp. 4569–71, on p. 4571.
19. H.-J. Rheinberger, *Experiment, Differenz, Schrift* (Marburg: Basilisken-Presse, 1992), p. 13.

20. Zahar, 'Logic of Discovery or Psychology of Invention?', p. 243.
21. F. Holmes, *Eighteenth-Century Chemistry as an Investigative Enterprise* (Berkeley, CA: Office for History of Science and Technology, University of California at Berkeley), 1989, p. xvi.
22. Rheinberger, *Experiment, Differenz, Schrift*.
23. P. Galison, *Image and logic: A Material Culture of Microphysics* (Chicago, IL: University of Chicago Press, 1997).
24. Harré, 'Recovering the Experiment'.
25. M. S. Morgan, 'Experiments without Material Intervention: Model Experiments, Virtual Experiments and Virtually Experiments', in H. Radder (ed.), *The Philosophy of Scientific Experimentation* (Pittsburgh, PA: University of Pittsburgh Press, 2003), pp. 216–35.
26. Steinle has elaborated the case of experiments on electromagnetism as an example of explorative experimentation (Steinle, *Explorative Experimente*, p. 129).
27. Heidelberger, 'Die Erweiterung der Wirklichkeit im Experiment', p. 72.
28. Ibid., p. 87.
29. Ibid., p. 88.
30. Hacking has made a similar statement with respect to 'observation without theory', thereby particularly objecting to Feyerabend's claim that observational reports are always charged with theoretical assumptions. Hacking argues that Feyerabend can only maintain this position given a wide and inaccurate use of terms, particularily of the term 'theory' and accordingly comments acidly: 'Feyerabend has shown us how not to talk about observation, speech, theory, habits, or reporting' (Hacking, *Representing and Intervening*, p. 175).
31. H.-J. Rheinberger, 'Plädoyer für eine Wissenschaftsgeschichte des Experiments', *Theory in Biosciences*, 116 (1997), pp. 11–31, on p. 25.
32. Rheinberger, *Experiment, Differenz, Schrift*, p. 72.
33. P. Galison, 'History, Philosophy, and the Central Metaphor', *Science in Context*, 2 (1988), pp. 197–212.
34. Galison, *How Experiments End*.
35. L. Fleck, *Denkstile und Tatsachen* (Berlin: Suhrkamp, 2011).
36. T. S. Kuhn, *The Structure of Scientific Revolutions* (Chicago, IL: University of Chicago Press, 1962) and *The Essential Tension*.
37. Harré, 'Recovering the Experiment', p. 353.
38. Indeed, Latour offers a largely abstract and semiotic description of how the gap is bridged between the representational and the material world. He speaks, for example, of a chain of representations, of translation processes, of distances and series and so on.
39. Harré, 'Recovering the Experiment', p. 366.
40. Ibid., pp. 376, 377.
41. Ibid., p. 366.
42. Ibid., p. 368.
43. Harré, 'The Materiality of Instruments in a Metaphysics for Experiments', p. 20.
44. Ibid., p. 27.
45. Harré, 'Recovering the Experiment', p. 376.
46. Ibid.
47. D. Baird, *Thing Knowledge: A Philosophy of Scientific Instruments* (Berkeley, CA: University of California Press, 2004); M. Carrier, *Wissenschaftstheorie: Zur Einführung* (Hamburg: Junius, 2006).

48. R. Laudan, 'Natural Alliance or Forced Marriage? Changing Relations between the Histories of Science and Technology', *Technology and Culture*, 36 (1995) (Supplement: Snapshots of a Discipline: Selected Proceedings from the Conference on Critical Problems and Research Frontiers in the History of Technology, Madison, Wisconsin, 30 October–3 November 1991), pp. 17–28, on pp. 17–18. She launches this discussion against the background of a debate about the power relations between the history of science and the history of technology by asking 'Natural Alliance or Forced Marriage? Changing Relations between the Histories of Science and Technology'.

49. Ibid., pp. 22–3.

50. Simon, 'Discovery, Invention, and Development', p. 4571.

51. W. Bainbridge and M. Roco (eds), *Converging Technologies for Improving Human Performance: Nanotechnology, Biotechnology, Information Technology and Cognitive Science*, NSF/DOC-Sponsored Report (Airlington, VA, prepublication online version, 2002), at http:// http://www.markswatson.com/NBIC.pdf [accessed 17 September 2013]; H. Etzkowitz, 'Innovation in Innovation: The Triple Helix of University-Industry-Government Relations', *Social Science Information*, 42 (2003), pp. 293–337; Nordmann (rapporteur), *Converging Technologies*; S. Miller, 'Science Confronts the Law', in W. Bainbridge and M. Roco (eds), *Managing Nano-bio-info-cogno Innovations: Converging Technologies in Society* (Dordrecht: Springer, 2006), pp. 279–84; B. R. Allenby and D. Rejeski, 'The Industrial Ecology of Emerging Technologies', *Journal of Industrial Ecology*, 12 (2008), pp. 267–9; J. Luce and L. Giorgi, 'Knowledge Politics and Converging Technologies', *Innovation: The European Journal of Social Science Research*, 22 (2009), pp. 1–5.

5 Experimentation in Lab and Field

1. B. Bensaude-Vincent, *Dans le laboratoire de Lavoisier* (Paris: Nathan, Monde en Poche, 1993); G. Wolters and J. G. Lennox (eds), *Concepts, Theories, and Rationality in the Biological Sciences* (Konstanz/Pittsburgh, PA: Universitätsverlag Konstanz/University of Pittsburgh Press, 1995); R. M. Burian, 'Comments on Complexity and Experimentation in Biology', *Philosophy of Science*, 64 (1997) (Supplement, Proceedings of the 1996 Biennial Meetings of the Philosophy of Science Association, Part II: Symposia Papers), pp. 279–91; Rheinberger, 'Plädoyer für eine Wissenschaftsgeschichte des Experiments'; R. Lange, *Experimentalwissenschaft Biologie. Methodische Grundlagen und Probleme einer technischen Wissenschaft vom Lebendigen* (Würzburg: Königshausen & Neumann, 1999); M. Weber, *Philosophy of Experimental Biology* (Cambridge: Cambridge University Press, 2005); K. Köchy and G. Schiemann (eds), 'Natur im Labor', *Philosophia naturalis*, 43 (2006), pp. 1–9.

2. Rudolf Kötter has noted with gentle irony that there is no 'physical field research' or 'field experimentation in physics', thus emphasizing that a physical experiment is always a technological enterprise, albeit aimed at registering and collecting physical entities and measurements. This argument comes close to Rheinberger's non-technical use of technology (R. Kötter, 'Experiment, Simulation und orientierender Versuch: Anmerkungen zur Experimentalkultur der Biowissenschaften', in C. Brunold, P. Balsiger, J. B. Bucher and C. Körner (eds), *Wald und CO2: Ergebnisse eines ökologischen Modellversuchs* (Bern: Haupt, 2001), pp. 41–50, on p. 42.

3. For instance D. R. Brothwell (ed.), *Environmental and Experimental Studies in the History of Archaeology*, Symposia of the Association for Environmental Archaeology No.

9, Roskilde, Denmark, 1988 (Oxford: Oxbow Books, 1990), pp. 1–23; R. Frodeman, *Geo-logic: Breaking Ground between Philosophy and the Earth Sciences* (Albany, NY: State University of New York Press, 2003); Gross, Hoffmann-Riem and Krohn, *Real-experimente*; W. Krohn, G. Böhme, W. van den Daele, R. Hohlfeld and W. Schäfer, *Finalization in Science: The Social Orientation of Scientific Progress* (Dordrecht: Reidel, 1983); Riecken, *Social Experimentation*; K. Shrader-Frechette and E. D. McCoy, *Method in Ecology: Strategies for Conservation* (Cambridge: Cambridge University Press, 1993).

4. Aquatic ecologist Walter Geller in his abstract for the ZIF workshop 'From Lab to Field: Transforming Research Practices', Bielefeld, 7 July 2007.

5. W. K. Michener, K. L. Bildstein, A. McKee, R. R. Parmenter, W. W. Hargrove, D. McClearn and M. Stromberg, 'Biological Field Stations: Research Legacies and Sites for Serendipity', *BioScience*, 59 (2009), pp. 300–10, on p. 308.

6. See the 'Field Studies: Network for the History and Sociology of Fieldwork and Scientific Expeditions' website, at http://www.fieldstudies.dk/ [accessed 20 August 2011].

7. Ibid.

8. The study tells the story of Danish traveller Henning Haslund-Christensen (Christel Braae, Anthropological Section, National Museum of Denmark).

9. Kohler, *Landscapes and Labscapes*, p. 6.

10. Carrier, *Wissenschaftstheorie: Zur Einführung*, p. 21.

11. K. Köchy, 'Lebewesen im Labor: Das Experiment in der Biologie', *Philosophia naturalis*, 43 (2006), pp. 74–110.

12. Ibid., p. 81.

13. Ibid., p. 83.

14. A. Rosenberg, *The Structure of Biological Science* (Cambridge: Cambridge University Press, 1985), pp. 27–8.

15. Kötter, 'Experiment, Simulation und orientierender Versuch'.

16. Weber, *Philosophy of Experimental Biology*.

17. Rheinberger, *Experiment, Differenz, Schrift*, pp. 73–80.

18. Köchy, 'Lebewesen im Labor. Das Experiment in der Biologie', p. 87.

19. Hacking, *Representing and Intervening*, p. 189.

20. Kenneth Waters has pointed out that the suitability of the material was 'essential for establishing *Drosophila* as a supermodel of experimental organisms' (K. Waters, 'How Practical Know-how Contextualizes Theoretical Knowledge: Exporting Causal Knowledge from Laboratory to Nature', *Philosophy of Science*, 75 (2008), pp. 707–19, on p. 716). The identification of suitable material was closely related to the way in which classical geneticists understood heredity, and thus to theory building. Moreover, Waters argues that the 'knowledge about what it took to make the experiments work provided an important source of information about the extent to which principles of the transmission theory might apply outside the laboratory' (ibid., p. 717).

21. U. Krohs, 'Wissenschaftstheoretische Rekonstruktionen', in U. Krohs and G. Toepfer (eds), *Philosophie der Biologie* (Frankfurt am Main: Suhrkamp, 2005), pp. 304–21; E. Mayr, *Das ist Biologie: Die Wissenschaft des Lebens* (Heidelberg: Spektrum, 2000); Rosenberg, *The Structure of Biological Science*.

22. Köchy, 'Lebewesen im Labor. Das Experiment in der Biologie', p. 98.

23. L. Daston, 'The Moral Economy of Science', *Osiris*, 10 (1995), pp. 3–24, on p. 24.

24. T. M. Porter, 'Objectivity and Authority: How French Engineers Reduced Public Utility to Numbers', *Poetics Today*, 2 (1991), pp. 245–65.

25. For instance, W. Kutschmann, *Der Naturwissenschaftler und sein Körper: Die Rolle der 'inneren Natur' in der experimentellen Naturwissenschaft der frühen Neuzeit* (Frankfurt am Main: Suhrkamp, 1986).

26. B. Latour and S. Woolgar, *Laboratory Life: The Social Construction of Scientific Facts* (Beverly Hills, CA: Sage, 1979); K. Knorr-Cetina, *The Manufacture of Knowledge: An Essay on the Constructivist and Contested Nature of Science* (Oxford: Pergamon Press, 1981); M. Lynch and S. Woolgar (eds), *Representation in Scientific Practice* (Dordrecht: Kluwer, 1990); A. Pickering, 'From Science as Knowledge to Science as Practice', in A. Pickering (ed.), *Science as Practice and Culture* (Chicago, IL: University of Chicago Press, 1992), pp. 1–29.

27. B. Latour, 'Give me a Laboratory and I Will Raise the World', in K. Knorr-Cetina and M. Mulkay (eds), *Science Observed* (London: Sage, 1983), pp. 141–70, on p. 141.

28. For instance D. G. Stern, 'The Practical Turn', in S. P. Turner and P. A. Roth (eds), *The Blackwell Guide to the Philosophy of the Social Sciences* (Oxford: Blackwell, 2003), pp. 185–206.

29. T. R. Schatzki and K. Knorr-Cetina (eds), *The Practice Turn in Contemporary Theory* (London: Routledge, 2001).

30. Knorr-Cetina, *The Manufacture of Knowledge*, p. viii.

31. P. Galison and E. Thompson (eds), *The Architecture of Science* (Cambridge, MA: MIT Press, 1999), p. 2.

32. Latour, 'Give me a Laboratory and I Will Raise the World', p. 160.

33. Ibid., p. 154.

34. K. Knorr-Cetina, 'The Couch, the Laboratory, and the Cathedrale: On the Relationship between Experiment and Laboratory in Science', in A. Pickering (ed.), *Science as Practice and Culture* (Chicago, IL: University of Chicago Press, 1992), pp. 113–38, on pp. 118–19.

35. This undermines the epistemic rupture between science and lay knowledge that was emphasized particularly by Gaston Bachelard: *La formation de l'esprit scientifique: Contribution à une psychoanalyse de la connaissance objective* (Paris: Vrin, 1938); *Epistemologie: Ausgewählte Texte* (Frankfurt am Main: Ullstein, 1974).

36. In Michel Callon's four model scheme describing the dynamic of science, competition is one model and science as rational knowledge, as socio-cultural practice, and as advanced translation are the other three (M. Callon, 'Four Models for the Dynamics of Science', in S. Jasanoff, G. E. Markle, J. C. Peterson and T. Pinch (eds), *Handbook of Science and Technology Studies* (Thousand Oaks, CA: Sage, 1995), pp. 29–63).

37. The relevance and presence of gestalt theory in scientific and philosophical discussions at the time might be supported by a comment of Ernst Cassirer, who attests that gestalt theory is a paradigm of a future interdisciplinarity between philosophy and the sciences (E. Cassirer, *Erkenntnis, Begriff, Kultur*, ed. and commentary by R. A. Bast (Hamburg: Felix Meiner, 1993).

38. L. Fleck, *Genesis and Development of a Scientific Fact*, trans. F. Bradley and T. J. Trenn (Chicago, IL: University of Chicago Press, 1979), p. 92.

39. It has recently been proposed that plasticity is a particular feature of technoscientific objects, one which distinguishes them from scientific objects. Unlike interpretive flexibility, plasticity is used here in a material idiom, indicating that objects of technoscientific interest might tend, in the end, 'to be immaterial to the point of becoming mere potentialities, totipotent entities' ('The Genesis and Ontology of Technoscientific

Objects', at http://www.philosophie.tu-darmstadt.de/media/philosophie___goto/text_1/gotoprojectoverview.pdf [accessed 21 August 2011]).

40. K. A. Rader, *Making Mice: Standardizing Animals for American Biomedical Research, 1900–1955* (Princeton, NJ: Princeton University Press, 2004).

41. K. Amann, 'Menschen, Mäuse und Fliegen: Eine wissenssoziologische Analyse der Transformation von Organismen in epistemische Objekte', *Zeitschrift für Soziologie*, 23 (1994), pp. 22–40, on p. 30.

42. A. Ferrari, *Genmaus & Co.: Gentechnisch veränderte Tiere in der Biomedizin* (Erlangen: Harald Fischer Verlag, 2008).

43. K. Knorr-Cetina, 'Die Manufaktur der Natur – oder: Die alterierten Naturen der Naturwissenschaft', in Institut für Wissenschafts- und Technikforschung (ed.), *Die 'Natur' der Natur* (Bielefeld: Institut für Wissenschafts- und Technikforschung, 1999), pp. 104–19.

44. Ibid., p. 105 (trans. AS).

45. K. Knorr-Cetina, *Epistemic Cultures: How the Sciences Make Knowledge* (Cambridge, MA: Harvard University Press, 1999), p. 26.

46. Bachelard, *Epistemologie: Ausgewählte Texte*, p. 20.

47. W.-E. Reif, 'Practice and Theory in Natural History: Domains and Epistemic Things', *Theory in Biosciences*, 118 (1999), pp. 161–74, on pp. 168–9.

48. Kohler, *Landscapes and Labscapes*.

49. A. Cunningham and P. Williams (eds), *The Laboratory Revolution in Medicine* (Cambridge: Cambridge University Press, 1992); S. Dierig, 'Engines for Experiment: Laboratory Revolution and Industrial Labor in the Nineteenth-Century City', *Osiris*, 18 (2003), pp. 116–34.

50. Kohler, *Landscapes and Labscapes*, p. 3.

51. Kohler claims that the question of why lab sciences are granted such a dominant place in modern societies cannot be considered as being answered in a satisfactory way as yet. Where do the labs derive their authority from – is it from the social context at the time, thus from the middle classes? Is it through alliances with 'mass public education', as Kohler speculatively suggests? Or is it because the scientific community offered a perfect blueprint for the transformation of 'the traditional business firm into the modern managerial leviathan'? (R. E. Kohler, 'Lab History: Reflections', *Isis*, 99 (2008), pp. 761–8, on p. 768).

52. Kohler, *Landscapes and Labscapes*, p. 3.

53. B. Bensaude-Vincent, 'Chemistry in the French Tradition of Philosophy of Science: Duhem, Meyerson, Metzger and Bachelard', *Studies in History and Philosophy of Science*, 36 (2005), pp. 627–48, on p. 642.

54. Dierig, 'Engines for Experiment', p. 118.

55. A. Schwarz, 'Modellierte Naturen und Raummodelle: Theoretische, ästhetische und strategische Eingriffe im Pariser Naturkundemuseum', *Berliner Schriften zur Museumskunde*, 22 (2006), pp. 155–64.

56. The History of the MNHN is discussed in detail in C. Limoges, 'The Development of the Muséum d'Histoire Naturelle of Paris, 1800–1914', in R. Fox and G. Weisz (eds), *The Organization of Science and Technology in France 1808–1914* (Cambridge: Cambridge University Press, 1980) pp. 211–40.

57. D. Outram, 'New Spaces in Natural History', in N. Jardine, J. A. Secord and E. Spary (eds), *Cultures of Natural History* (Cambridge: Cambridge University Press, 1996), pp. 249–479.

58. Kohler, *Landscapes and Labscapes*, p. 2.

59. Livingstone, *Putting Science in its Place*.
60. Ibid., p. 40.
61. Ibid., p. 41.

6 Exploring the Field Ideal

1. P. Bliss, *The Heart of Emerson's Journals* (New York: Dover Publications, 1995).
2. H. Lacey, 'On the Aims and Responsibilities of Science', *Principia*, 11 (2007), pp. 45–62.
3. Kohler, *Landscapes and Labscapes*, p. 191.
4. L. Daston and E. Lunbeck, 'Introduction', in L. Daston and E. Lunbeck (eds), *Histories of Scientific Observation* (Chicago, IL: University of Chicago Press, 2011), pp. 1–9, on p. 3.
5. For example, with regard to the testability of theory, Rosenberg, *The Structure of Biological Science*, p. 27.
6. Kohler, *Landscapes and Labscapes*, p. 193.
7. M. Foucault, 'Of Other Spaces', *Diacritics*, 16 (1986), pp. 22–7.
8. J. Malpas, 'Finding Place: Spatiality, Locality, and Subjectivity', in A. Light and J. M. Smith (eds), *Philosophies of Place* (Lanham: Rowman and Littlefield, 1998), pp. 21–44, on p. 21.
9. D. Massey, *For Space* (London: Sage, 2005), p. 15.
10. In the 1980s this was identified as a 'geographical turn' – to be dismantled in the 1990s as a myth (J. Duncan, 'The Hidden Geographies of Social Sciences and the Myth of the Geographical Turn', *Environment and Planning D: Society and Space*, 13 (1995), pp. 379–80).
11. R. Barthes, *Mythologies* (New York: Hill and Wang, 1972).
12. '[T]he enjoyment of being enclosed reaches its paroxysm when, from the bosom of this unbroken inwardness, it is possible to watch, through a large window-pane, the outside vagueness of the waters, and thus define, in a single act, the inside by its opposite' (ibid., p. 66).
13. M. Gibbons, C. Limoges, H. Nowotny, S. Schwartzmann, P. Scott and M. Trow, *The New Production of Knowledge: The Dynamics of Science and Research in Contemporary Societies* (London: Sage Publications, 1994).
14. In some of the smaller cantons in Switzerland, voting in public elections still takes place in the marketplace (e.g. Appenzell).
15. An argument about the urban place as a socio-political spot of enabling engagement and interaction and the role of architecture to afford those built environments was put forward by Ludger Schwarte in his book on *Philosophie der Architektur* (Munich: Fink, 2009). He offers a reconstruction of the French Revolution with a special emphasis on the siting of events, that is, on which place what happened.
16. The following selection of edited volumes gives an idea of the presence of place and the diversity of perspectives: I. Billick and M. V. Price (eds), *The Ecology of Place: Contributions of Place-based Research to Ecological Understanding* (Chicago, IL: University of Chicago Press, 2010); A. Light and J. M. Smith (eds), *Philosophies of Place (Philosophy and Geography III)* (Lanham, MD: Rowman and Littlefied Publishers, 1998); P. Descola and G. Palsson, *Nature and Society: Anthropological Perspectives* (London: Routledge, 1996).
17. Malpas, 'Finding Place', p. 36.
18. Adopting a sociological perspective, Martina Löw developed a conception in which space is considered to be a 'relational disposition/regime [*(An)Ordnung*] of living beings

and social goods' (*Raumsoziologie* (Frankfurt am Main: Suhrkamp, 2001), p. 160). With this emphasis on relationships creating connectedness, her conception agrees with a theoretical approach in early geography, social anthropology and ecology. Gibson's ecological psychology, for example, also refers to this motif, in that the physical structures of the environment and of the organism are thought to be mutually constraining and complementary (Gibson, *The Ecological Approach to Visual Perception*).

19. In his book *On Settling* (Princeton, NJ: Princeton University Press, 2012), philosopher Robert E. Goodin follows a congenial line of argument about connectedness. He explores different 'modes of settling', offering a phenomenology of settling that is more interested in the practices represented than in the phrasing of these practices.

20. A. Schwarz, 'Baron Jakob von Uexküll: Das Experiment als Ordnungsprinzip in der Biologie', in A. Schwarz and A. Nordmann (eds), *Das bunte Gewand der Theorie* (Freiburg: Alber, 2009), pp. 207–34.

21. A. Naess, 'The Shallow and the Deep, the Long-range Ecology Movement: A Summary', *Inquiry*, 16 (1973), pp. 95–100.

22. G. Böhme and G. Schiemann, *Phänomenologie der Natur* (Frankfurt am Main: Suhrkamp, 1997).

23. C. Rehmann-Sutter, 'An Introduction to Places', *Worldviews: Environment, Culture, Religion*, 2 (1998), pp. 171–7.

24. There is ongoing debate about collaboration as opposed to cooperation in the field of education and learning and in computer science, among others. For instance, the solution to an open-source problem may arise out of a cooperative effort – that is, it may be worked out by individual people with a common goal who each contribute time and expertise to finding it. This is juxtaposed to collaborative work on solving a problem, which might also happen online, but those involved are more intensively engaged, not least in discussions with one another. The benefits of the transformation from a cooperative to a collaborative relationship can be seen mainly in the support expressed for sustained conversations and in the provision of an institutionalized structure for weaving together multiple perspectives.

25. S. L. Star and J. R. Griesemer, 'Institutional Ecology, "Translations" and Boundary Objects: Amateurs and Professionals in Berkeley's Museum of Vertebrate Zoology, 1907–39', *Social Studies of Science*, 19 (1989), pp. 387–420, on pp. 388–9, 413.

26. B. Carr, 'Popper's Third World', *Philosophical Quarterly*, 27 (1977), p. 219.

27. Star and Griesemer, 'Institutional Ecology, "Translations" and Boundary Objects'.

28. The antagonist terms diographic and nomothetic had already been introduced in nineteenth-century philosophy of science (W. Windelband, *Präludien: Aufsätze und Reden zur Einleitung in die Philosophie* (Tübingen: Mohr, 1884)).

29. U. Eisel, 'Triumph des Lebens: Der Sieg christlicher Wissenschaft über den Tod in Arkadien', *Urbs et Regio Sonderband*, 65 (1997), pp. 39–160, on p. 104.

30. The identification of nomothetic methodology predominantly with the natural sciences was not intended by Windelband. Indeed, he regretted the way in which boundaries had been drawn around different realms of objects, describing these divisions as 'unfortunate' (W. Windelband, *Geschichte und Naturwissenschaft: Rede zum Antritt des Rektorats der Kaiser-Wilhelm-Universität Strassburg; gehalten am 1. Mai 1894*, 2nd edn (Strasbourg: Heitz, 1900), p. 26).

31. Cunningham and Williams, *The Laboratory Revolution in Medicine*, p. 6.

32. Humboldt (1823), in M. Bowen, *Empiricism and Geographical Thought: From Francis Bacon to Alexander von Humboldt* (Cambridge: Cambridge University Press, 1981), p. 233.

33. C. Kwa, *Styles of Knowing: A New History of Science from Ancient Times to the Present* (Pittsburgh, PA: University of Pittsburgh Press, 2011), pp. 427–8.

34. P. W. Balsiger, 'Überlegungen und Bemerkungen hinsichtlich einer Methodologie inter-disziplinärer Wissenschaftspraxis', in P. W. Balsiger, R. Defila and A. di Giulio (eds), *Ökologie und Interdisziplinarität: eine Beziehung mit Zukunft?* (Basel: Birkhäuser, 1996), pp. 73–85; A. Valsangiacomo, *Die Natur der Ökologie* (Zurich: Hochschulverlag ETH Zürich, 1998).

35. Some of the following examples are adopted from Schwarz and Krohn, 'Experimenting with the Concept of Experiment'.

36. The institute was renamed the Max Planck Institute for Evolutionary Biology in March 2007.

37. E. P. Odum, 'The Mesocosm', *BioScience*, 24 (1984), pp. 558–62, on p. 558.

38. 'The originality of mesocosm experimentation is mainly based on the combination of ecological realism, achieved by introduction of the basic components of natural ecosys-tems, and facilitated access to a number of physicochemical, biological, and toxicological parameters that can be controlled to some extent ... Each natural ecosystem is unique because its structure and function mainly depend on local factors. Therefore, there is a conceptual opposition between realism and replicability when applied to mesocosms' (T. Caquet, L. Lagadic and S. R. Sheffield, 'Mesocosms in Ecotoxicology. Outdoor Aquatic Systems', *Reviews of Environmental Contamination and Toxicology*, 165 (2000), pp. 1–38, on p. 1).

39. The website http://mesocosm.eu offers a sort of anthology of mesocosms, providing information about mesocosm facilities worldwide.

40. E. Naumann, 'Die Rohkultur des Heleoplanktons', in E. Abderhalden (ed.), *Handbuch der biologischen Arbeitsmethoden. Methoden der Süsswasserbiologie*, part 2 (Berlin: Urban und Schwarzenberg, 1925), p. 283.

41. H. Dietz and T. Steinlein, 'Recent Advances in Understanding Plant Invasions', *Progress in Botany*, 65 (2004), pp. 539–73.

42. This was discussed in more detail with a special emphasis on climate change research in a paper by geographer-climatologist Stefan Brönnimann and philosopher Gertrude Hirsch Hadorn, 'Lessons for Science from the "Year without a Summer" of 1816: What does it Take for Science to Respond to Climatic Change?', *Gaia*, 22 (2013), pp. 169–73.

43. Transregional Collaborative Research Centre SFB/TRR 38: 'Structures and processes of the initial ecosystem development phase in an artificial water catchment', at http://www.tu-cottbus.de/sfb_trr/eng/index.htm [accessed 9 July 2011].

44. A more detailed narrative of this attractive technoscientific object and an analysis of the epistemic-ontological conditions of this real-world simulation is offered in A. Schwarz, 'In the Beginning, Man Created ...: Narrating the Drama of an Emerging Ecosystem', in B. Bensaude-Vincent, S. Loeve, A. Nordmann and A. Schwarz (eds), *Attractive Objects* (Pittsburgh, PA: Pittsburgh University Press, forthcoming).

45. Krohn and van den Daele, 'Science as an Agent of Change', p. 195.

46. G. D. Rose, 'Social Experiments in Innovative Environmental Management: The Emer-gence of Ecotechnology' (PhD thesis, University of Waterloo, 2003), p. 165.

47. Gross, Hoffmann-Riem and Krohn, *Realexperimente*.

48. Hoffmann-Riem in ibid., p. 137.

7 Stretching the Baconian Contract – but How Far?

1.   Bacon, *Neues Organon*, ed. Krohn, vol. 1, a 114.

2.   Gibbons et al., *The New Production of Knowledge*; L. Schäfer, *Das Bacon-Projekt: Von der Erkenntnis, Nutzung und Schonung der Natur* (Frankfurt am Main: Suhrkamp, 1999).

3.   M. Serres, *Retour au contrat naturel* (Paris: Bibliothèque Nationale de France, 2000).

4.   This argument was discussed in more detail in the paragraph 'Lab-field Co-production' in Chapter 5 'Experimentation in Lab and Field'.

5.   The bouillon cube was only put on the market in 1908 by Maggi, a Swiss firm founded by Julius Maggi, but the idea and formulation of a concentrated meat extract had already been invented around 1840 by Justus von Liebig. 'Experimenting with calories' refers to the studies on the heat value of nutrients and the calculation of energy that was expressed in calories. From the last quarter of the nineteenth century on, the efficiency of blue-collar workers was measured in terms of their calorie balance – 'Taylorism in action' (C. Jou, 'Counting Calories', *Chemical Heritage*, 29 (2011), pp. 27–31, on p. 30). The deployment of calories as a relentless means to control the shape of women followed a little later.

6.   Wilhelm Einsele, researcher at the Institut für Seenforschung, Langenargen, started his first 'pre-experiment' in summer 1937, adding 1400kg of *Superphosphat* to the lake. The standard fertilizer came in small pellets and was donated by the German Superphosphat-Gesellschaft (Berlin). The phosphate pellets were strewn from a boat that went back and forth across the lake systematically. It was hoped that the experiment would help to establish what the limiting factors are in so-called lake production. As it turned out, Einsele was able to present the first fully fledged 'image' of the phosphorus cycle (and a correlation with sources of ferric oxides) based on mass balances (W. Einsele, 'Die Umsetzung von zugeführtem, anorganischen Phosphat im eutrophen See und ihre Rückwirkung auf seinen Gesamthaushalt', *Zeitschrift für Fischerei und deren Hilfswissenschaften*, 39 (1941), pp. 407–88). Since then, every limnological textbook gives phosphorous as the limiting factor of a lake system. The problem, originally developed in an applied context, had turned into a theoretical model.

7.   'Think like an Ecosystem' is the title of a paper that argues for the implementation of a 'living system paradigm' in participatory planning. It puts forward a model of participatory ecological design that involves exploring the role of active stakeholder participation in 'planning for sustainability' (J. Tippett, '"Think Like an Ecosystem": Embedding a Living System Paradigm into Participatory Planning', *Systemic Practice and Action Research*, 17 (2005), pp. 603–22).

8.   This was elaborated in more detail in Schwarz and Krohn, 'Experimenting with the Concept of Experiment', to which parts of this chapter refer.

9.   See Chapter 5 'Experimentation in Lab and Field'.

10. M. Carrier, 'Theories for Use: On the Bearing of Basic Science on Practical Problems', in *EPSA07: 1st Conference of the European Philosophy of Science Association* (Madrid, 15–17 November 2007), p. 32, at http://philsci-archive.pitt.edu/id/eprint/3690 [accessed 21 December 2013].

11. N. Rupke (ed.), *Vivisection in Historical Perspective* (London: Routledge, 1990); A. Guerrini, 'Animal Experiments and Antivivisection Debates in the 1820s', in C. Knellwolf and J. Goodall (eds), *Frankenstein's Science: Experimentation and Discovery in Romantic Culture, 1780–1830* (Aldershot and Burlington, VT: Ashgate, 2008), pp. 71–86.

12. J. Moreno, *Undue Risk: Secret State Experiments on Humans* (London: Routledge, 2000); A. Goliszek, *In the Name of Science: A History of Secret Programs, Medical Research, and Human Experimentation* (New York: St Martin's Press, 2003). A good example for the misleading authority of experimental truth is the famous Milgram experiments. They seemed to justify the character of humans as having an untamable disposition to violoence and cynism. Recently, Hans Bernhard Schmid has pointed out that this is an interpretation that rather owes to the biased experimental setting (and also, of course, to Hannah Arendt's dictum of the banality of evil, which was quite topical at the time) than to the violent disposition of the experimental objects – or even of humankind. In Milgram's learning experiments, the subjects were split into two groups, teachers and learners, in which the teachers were allowed to apply an electric shock to the learners if the latter failed to learn correctly. The teachers and their learners became involved in a rising scale of violence, with the alarming outcome that 62 per cent of the participants in the teacher role were prepared to apply electric shocks to their victims to the point of death. In his detailed study Schmid shows convincingly that this result owes rather to the methodology than to the moral laxity of the research subjects. He argues that the experimental setting produced a paradox situation in that no real discourse was possible because both parties had to react in a prescribed manner – as a kind of automaton. Stanley Milgram chose the same setting and theories to conduct his behavourist experiments with rats (whose adequacy is likewise debatable). Schmid concludes that it was not the research subjects who were lacking in moral integrity, but that it was the experiment itself that had no integrity (H. B. Schmid, *Moralische Integrität* (Frankfurt am Main: Suhrkamp, 2011)).

13. J. Dewey, *Context and Thought* (Berkeley, CA: University of California Publications in Philosophy, 1929), p. 133.

14. K. Popper, *The Open Society and its Enemies: The Spell of Plato* (Princeton, NJ: Princeton University Press, 1945).

15. Ibid., p. 158.

16. Ibid.

17. Ibid., p. 159.

18. Ibid., p. 162.

19. A. W. Small, 'The Future of Sociology', *Publications of the American Sociological Society*, 16 (1921), pp. 174–93, on p. 187.

20. T. Gieryn, 'City as Truth-Spot: Laboratories and Field-Sites in Urban Studies', *Social Studies of Science*, 36 (2006), pp. 5–38, on p. 8.

21. D. T. Campbell and J. C. Stanley, 'Experimental and Quasi-experimental Designs for Research on Teaching', in N. L. Gage (ed.), *Handbook of Research and Teaching* (Chicago, IL: Rand McNally, 1963), pp. 171–246; Campbell, 'Reforms as Experiments'.

22. C. F. Sabel and J. Zeitlin, 'Learning from Difference: The New Architecture of Experimentalist Governance in the EU', *European Law Journal*, 13 (2007), pp. 271–327; M. Yuval and J. Lezaun, 'Regulatory Experiments: Genetically Modified Crops and Financial Derivatives on Trial', *Science and Public Policy*, 33 (2006), pp. 179–90; Felt et al., *Taking European Knowledge Society Seriously*, pp. 66–7.

23. See note above in the paragraph on 'Field Experimentation in Society'.

24. B. A. Minteer and B. P. Taylor (eds), *Democracy and the Claims of Nature* (New York: Rowman and Littlefield, 2002).

8 About the Epistemology and Culture of Borders/Boundaries

1. K. Parsons (ed.), *The Science Wars: Debating Scientific Knowledge and Technology* (New York: Prometheus Books, 2003), pp. 14–15.

2. In 2010 the German Federal Ministry of Education and Research (BMBF) spent about 6 per cent of its budget for research and development on risk and 'accompanying' research. This is more than any other major industrialized nation (compared to the USA, UK or Japan) and thus gives an indication of the relative lack of importance attached to this research. Germany's NanoKommission 2011 Report recommends that the subsidies for 'risk and accompanying research' should be 'significantly increased' (Bundesministerium für Umwelt und Reaktorsicherheit (BMU) (ed.), *Verantwortlicher Umgang mit Nano-technologien. Bericht und Empfehlungen der NanoKommission 2011*, text written by A. Grobe, ed. W.-M. Catenhusen (Berlin: BMU, 2010), p. 12).

3. NSF/EC Workshop nano2, Hamburg, June 2010 (see also M. C. Roco, B. Harthorn, D. Guston and P. Shapira, 'Innovative and Responsible Governance of Nanotechnology for Societal Development', in M. C. Roco, C. Mirkin and M. Hersam (eds) *Nanotechnology Research Directions for Societal Needs in 2020* (Berlin: Springer, 2011), pp. 561–617, on pp. 561–2).

4. See, for example, a provision in the German constitution that declares concern for future generations and for the conditions of animal life to be a national objective (27 October 1994): 'Der Staat schützt auch in Verantwortung für die künftigen Generationen die natürlichen Lebensgrundlagen und die Tiere im Rahmen der verfassungsmäßigen Ordnung durch die Gesetzgebung und nach Maßgabe von Gesetz und Recht durch die vollziehende Gewalt und die Rechtsprechung' (GG article 20a).

5. Simon Schaffer has pointed out that the powers of disciplinarity 'highlight the intriguing relation between discipline as a form of knowledge organization, as a form of surveillance and as a form of exercise'. From this he concludes that 'the discourse of interdisciplinarity must change its historiography of Eurocentric and monolithic disciplinarity and must begin to explore the historical geography of exotic indisciplines' (manuscript of a talk given at TU Darmstadt in October 2010).

6. I. Kant, 'Beschluß von der Grenzbestimmung der reinen Vernunft §57', in *Prolegomena zu einer jeden künftigen Metaphysik, die als Wissenschaft wird auftreten können*, ed. K. Vorländer (Leipzig: Felix Meiner, 1920), pp. 120–30.

7. T. Gieryn, 'Boundary Work and the Demarcation of Science from Non-science: Strains and Interests in Professional Ideologies of Scientists', *American Sociological Review*, 48 (1983), pp. 781–95; 'Boundary Object': Star and Griesemer, 'Institutional Ecology, "Translations" and Boundary Objects'; 'Trading Zone': P. Galison and D. J. Stump (eds), *The Disunity of Science: Boundaries,Ccontexts, and Power* (Stanford, CA: Stanford University Press, 1996); 'border zone': Kohler, *Landscapes and Labscapes*.

8. Wolfgang van der Daele, sociologist and environmentalist, cited in a conference report by W. Rammert, 'Eine Soziologie, als ob Natur nicht zählen würde', at http://www.ts.tu-berlin.de/fileadmin/fg226/Rammert/articles/Soziologie_der_Natur.html [accessed 4 December 2013].

9. K. Ott, *Aufbruch und Wandel: Regelwerke für einen Green New Deal* (2010), at http://www.bfn.de/fileadmin/MDB/documents/ina/vortraege/2010_Sommerakademie-Konrad-Ott.pdf [accessed 4 December 2013].

10. H. Sukopp, *Rückeroberung? Natur im Großstadtbereich* (Wien: Picus, 2003), p. 31.

11. L. Wittgenstein, *Philosophical Investigations* (Oxford: Basil Blackwell, 1958), p. 47 (remark 109).

12. This very personal affirmation of love is – perhaps precisely because it comes across as rather pathos-laden, appearing quite abruptly in a scholarly work on philosophy – an intriguing example of the operational character of language. In this case it turns a declaration of love, usually a very personal act, into a philosophically rendered public vow of allegiance, akin to a contract (A. Nordmann, *Wittgenstein's Tractatus: An Introduction* (Cambridge: Cambridge University Press, 2005), pp. 205–6).

13. 'Fläche ist die Grenze des körperlichen Raumes, indessen doch selbst ein Raum, Linie ein Raum, der die Grenze der Fläche ist, Punkt die Grenze der Linie, aber doch noch immer ein Ort im Raume' (I. Kant, 'Beschluß von der Grenzbestimmung der reinen Vernunft §57', p. 126).

14. B. Latour, 'From the World of Science to the World of Research?', *American Association for the Advancement of Science*, 280 (1998), pp. 208–9.

15. This is, of course, a rather crude characterization for the sake of contrast in the following discussion. The arguments on scientific realism are also of relevance in Chapter 4, the section 'A Case Study: Stepping Outside – Victorian Tidal Studies' and Chapter 5 'Experimentation in Lab and Field'. For a more comprehensive discussion, see for instance Hacking's 'Entity Realism' (Hacking, *Representing and Intervening*) and Chakravartty's 'Structural Realism' (A. Chakravartty, 'Semirealism', *Studies in History and Philosophy of Science*, 29 (1998), pp. 391–408), the latter treating 'entity' and 'structural' realism as two forms of a 'semi-realism'.

16. Kohler, *Landscapes and Labscapes*, p. 19.

17. See, for instance, C. H. Shofield (ed.), *Global Boundaries* (London: Routledge 1994), pp. 1–15.

18. T. Gieryn, 'Boundaries of Science', in S. Jasanoff (ed.), *Handbook of Science and Technology Studies* (London: Sage Publications 1995), pp. 393–443; T. Gieryn, *Cultural Boundaries of Science: Credibility on the Line* (Chicago, IL: University of Chicago Press, 1999).

19. Star and Griesemer, 'Institutional Ecology, "Translations" and Boundary Objects'.

20. Galison and Thompson (eds), *The Architecture of Science*.

21. Kohler, *Landscapes and Labscapes*.

22. C. L. Palmer, *Work at the Boundaries of Science: Information and the Interdisciplinary Research Process* (Dordrecht: Kluwer Academic Publishers, 2001).

23. Star and Griesemer, 'Institutional Ecology, "Translations" and Boundary Objects', pp. 396, 393.

24. Galison and Stump (eds), *The Disunity of Science*, p. 119.

25. Kohler, *Landscapes and Labscapes*, p. 11.

26. O. Martinez, 'The Dynamics of Border Interaction: New Approaches to Border Analysis', in C. H. Shofield (ed.), *Global Boundaries* (London: Routledge, 1993), pp. 1–15, on p. 5.

27. Ibid.

9 Excursus: 'Bridging Science' or 'Problem-Based Science'?

1. A. Thienemann, *Die Bedeutung der Limnologie für die Kultur der Gegenwart* (Stuttgart: Schweizerbart'sche Verlagsbuchhandlung, 1935), p. 20.

2. Ibid., p. 19.

3. W. Burkamp, *Die Struktur der Ganzheiten* (Berlin: Junker und Dünnhaup, 1929), p. 361: 'Die Gefahr der Abgeschlossenheit von Wissenschaften wird durch die zahlreichen sich zwischengruppierenden Rand- und Brückenwissenschaften gemildert'.

4. Ibid., p. 365.

5. Ibid., p. 366.

6. Compare also recent reflections on interdisciplinarity and problem-oriented research claiming a type of 'problem-oriented interdisciplinarity' and trying to carve out an operational and sound classification of inter/transdisciplinarity 'to shed some light on the vague notion of "problem"' (J. C. Schmidt, 'What is a Problem? On Problem-oriented Interdisciplinarity', *Poiesis and Praxis*, 7 (2011), pp. 249–74, on p. 249). This effort leads us to conclude that interdisciplinary research differs mainly from disciplinary research in that the latter pursues (research) programmes whereas the former defines projects that 'can't be solved but to clarify the problem and back-translate that which makes a problem a problem' (ibid., p. 270). Burkamp's semantics might offer a complementary perspective while pointing to the character of relations (epistemological and institutional) in problem-based activities.

7. A. Thienemann, 'Zwecke und Ziele der Internationalen Vereinigung für theoretische und angewandte Limnologie', *Verhandlungen der Internationalen Vereinigung für theoretische und angewandte Limnologie*, 1 (1923), pp. 1–5, on p. 3.

8. Accordingly, in 1956 the Department of Agriculture and Horticulture at Humboldt University, Berlin conferred on him the degree of Dr. agrar. h.c.

9. A. Thienemann, 'Vom Wesen der Ökologie', *Biologia Generalis*, 3/4 (1942) (special edition), pp. 312–31, on p. 324.

10. G. H. Schwabe, 'August Thienemann in Memoriam', *Oikos*, 12 (1961), pp. 310–16, on p. 316.

11. A. Thienemann, 'Der Nahrungskreislauf im Wasser', *Zoologischer Anzeiger*, Suppl. 2 (1927), pp. 29–79, on p. 33.

12. Kohler, *Landscapes and Labscapes*. The bridge metaphor, too, is taken up time and again as a means of characterizing ecology. The tradition and tenability of the semantics of bridging are reconstructed by H. W. Ingensiep, 'Brückenschläge ☒ zur Sprache der Ökologie', in B. Busch (ed.), *Jetzt ist die Landschaft ein Katalog voller Wörter: Beiträge zur Sprache der Ökologie* (Göttingen: Wallstein, 2007), pp. 128–37; and T. Potthast, '"Ökologie" als Brücke zwischen Wissen und Moral der Natur?', in B. Busch (ed.), *Jetzt ist die Landschaft ein Katalog voller Wörter: Beiträge zur Sprache der Ökologie* (Göttingen: Wallstein, 2007), pp. 138–45.

13. Another aspect of this bourgeois self-image of ecology is Thienemann's emphasis on the personality of the researcher: 'Every scientific study that rises above the average has a markedly personal, that is to say, a subjective touch; and it is often this that makes such a study especially interesting. This is why I think we can only really fully appreciate the work of our scientific colleagues when we know them personally' (Thienemann, 'Zwecke und Ziele der Internationalen Vereinigung für theoretische und angewandte Limnologie', pp. 4–5).

14. A. Thienemann, *Grundzüge einer allgemeinen Ökologie* (Stuttgart: Schweizerbart'sche Verlagsbuchhandlung, 1939), p. 12.

15. A. Thienemann, 'Vom Gebrauch und vom Mißbrauch der Gewässer in einem Kulturlande', *Archiv für Hydrobiologie*, 45 (1951), pp. 557–83, on p. 580.

16. A. Thienemann, 'Wasser – Das Blut der Erde', in F. Oppenberg (ed.), *Handbuch der Schutzgemeinschaft Deutscher Wald: Uns ruft der Wald* (Rheinhausen: Verlagsanstalt Rheinhausen, 1954), pp. 45–9, on p. 49.

17. Thienemann, 'Vom Wesen der Ökologie', p. 326.

18. E. Wasmund, 'Wissenschaftsprovinzen', *Deutsche Rundschau*, 52 (1926), pp. 243–53, on p. 245.

19. Joachim Radkau draws attention to the fact that during the 1930s and 1940s proponents of nature and environmental conservation held very different positions regarding 'nature', some of which are more adequately described by reference to a polycratic rather than a totalitarian model (J. Radkau, 'Naturschutz und Nationalsozialismus – wo ist das Problem?', in J. Radkau and F. Uekötter (eds), *Naturschutz und Nationalsozialismus* (Frankfurt am Main: Campus, 2003), p. 43).

10 Ecotechnology Complements Ecoscience: Probing a Framework

1. This a quote from a four-page statement put out by the steering committee of SAGUF, the Swiss Academic Society for Environmental Research and Ecology (G. H. Hadorn, U. Kunz and M. Maibach, 'Umweltforschung: Nice to Have or Need to Have?', *Gaia*, 13 (2004), pp. 70–3).

2. The philosophical accounts provided by Bataille and by Arendt speak rather of 'exuberance' than of 'abundance'; this will be discussed in more detail below in the section dealing with folk theories about 'eco' and 'nano' in Chapter 11, and mainly in Chapter 13 'Political Economy of Experiments'.

3. For a more detailed discussion of the four-field scheme proposed by Donald Stokes, see Figure 7.1 along with the explanation in the 'Experimentation and Innovation in Society' section in Chapter 7.

4. A special issue of the *Journal of Industrial Ecology* asked in 2008 'what industrial ecology has to offer for an emerging technology and what that emergence has to say for the development of industrial ecology' (R. Clift and S. Lloyd, 'Nanotechnology. A New Organism in the Industrial Ecosystem?', *Journal of Industrial Ecology*, 12 (2008), pp. 259–62, on p. 259).

5. A more detailed discussion of Whewell as an advocate of sciences based on rules and representations but not necessarily on laws can be found in the section headed 'Finding Conceptions by Constructive Roles' in Chapter 3 'Whewell's Innovation through Renewal'.

6. Some of these theories and models originated in an applied context but evolved further and eventually came to conform to the ideal of fundamental research.

7. This juxtaposition of 'science' and 'technoscience' follows a rather familiar scheme (Carrier, *Wissenschaftstheorie: Zur Einführung*; N. Cartwright, *The Dappled World: A Study of the Boundaries of Science* (Cambridge: Cambridge University Press, 1999); Hacking, *Representing and Intervening*; A. Nordmann, 'Collapse of Distance: Epistemic Strategies of Science and Technoscience', *Danish Yearbook of Philosophy*, 41 (2007), pp. 7–34)). However, what is unusual is its application to the field of ecological research, where ecotechnology is in keeping with a technoscientific framework and ecoscience with science conceived in a traditional sense. However, whether ecoscience can be understood from the perspective of traditional philosophy of science is another question.

8. Donald Worster offered a canonical work on *Nature's Economy: A History of Ecological Ideas* (Cambridge: Cambridge University Press, 1985). As to Haeckel's 'Oekologie',

none of the definitions offered by him denoted an existing research programme, nor was his aim to develop such a programme or even an institutional setting for ecological research. The term primarily filled a gap that had existed within the disciplinary system of zoology, namely, that of 'external physiology'. As such it reflects Haeckel's search for a systematic ordering of the study of living beings – an endeavour that was perfectly in line with contemporary efforts to find a general system of biology. To be sure, the antecedents of an ecological idea in the sense of a modern science and in contrast to natural history did already exist. Such ideas were present, for example, in Alexander von Humboldt's 'physiognomic system of plant forms', Alfred R. Wallace's 'geography' of animal species, and Charles Darwin's 'entangled bank'; they can also be found in Louis Agassiz's studies on lakes and oceans. A more extended history of ecology's first tentative steps is presented in A. Schwarz, 'Etymology and Original Sources of the Term "Ecology"', in A. Schwarz and K. Jax (eds), *Ecology Revisited: Reflecting on Concepts, Advancing Science* (Dordrecht: Springer, 2011), pp. 145–8, and also in A. Schwarz and K. Jax, 'Early Ecology in the German-speaking World through WWII', in A. Schwarz and K. Jax (eds), *Ecology Revisited: Reflecting on Concepts, Advancing Science* (Dordrecht: Springer, 2011), pp. 231–76.

9. Thus this method is diametrically opposed to the nomothetic methodology (e.g. of physics), which is guided by the idea of lawfulness and regularities and is most often identified with the deductive method. For more details, see the sectino headed 'Placing Place' in Chapter 6 'Exploring the Field Ideal'.

10. A detailed historical study on the age of ecology was delivered by J. Radkau, *Die Ära der Ökologie: Eine Weltgeschichte* (Munich: Beck, 2011).

11. The conception of post-normal science was proposed by Funtowicz and Ravetz in 1994 and places special emphasis on environmental matters (S. Funtowicz and J. Ravetz, 'Uncertainty, Complexity and Post-normal Science', *Environmental Toxicology and Chemistry*, 13 (1994), pp. 1881–5).

12. E. Becker, 'Transformations of Social and Ecological Issues into Transdisciplinary Research', in UNESCO/EOLSS Publishers (ed.), *Knowledge for Sustainable Development: An Insight into the Encyclopedia of Life Support Systems*, 3 vols (Paris and Oxford: EOLSS Publishers, 2002), vol. 3, pp. 949–63, on p. 949. Further aspects of transdisciplinary research with its 'quandaries' and 'hybridities' have been proposed by a number of scholars, such as F. Wickson, 'Transdisciplinary Research: Characteristics, Quandaries and Qualities', *Nature*, 456 (2008), p. 29. An overview is given in K. Pezzoli, 'Sustainable Development Literature: A Transdisciplinary Bibliography', *Journal of Environmental Planning and Management*, 40 (1997), pp. 575–601, and also in C. Pohl and G. H. Hadorn (eds), *Principles for Designing Transdisciplinary Research* (Munich: Oekom-Verlag, 2007).

13. This observation is confirmed in the literature, and ecology's plurality has become almost a commonplace. This aspect, along with the way it links in with the more general debate about plurality in the philosophy of science, is discussed in more detail in Schwarz, 'Etymology and Original Sources of the Term "Ecology"', upon which parts of this chapter draw.

14. Nancy Cartwright has shown that, if anything, the search for explanatory unity detracts from the search for truth (Cartwright, *The Dappled World*).

15. Lakatos, *The Methodology of Scientific Research Programmes*, p. 135.

16. H. E. Longino, 'Theoretical Pluralism and the Scientific Study of Behaviour', in S. H. Kellert, H. E. Longino and C. K. Waters (eds), *Scientific Pluralism: Minnesota Studies in*

the Philosophy of Science (Minneapolis, MN and London: University of Minnesota Press, 2006), pp. 102–31, on p. 127.

17. R. Carnap, "Meaning and Necessity: A Study in Semantics and Modal Logic", *Revue Internationale de Philosophie*, 4 (1950; reprinted by University of Chicago Press, 1956), pp. 20–40, on p. 40.

18. There has been a long-standing and not always constructive debate in the philosophy of science about whether laws can exist in biology at all or whether they are confined solely to physics (or physics and chemistry). This is, of course, mainly a question of how 'laws' in the scientific realm are to be understood. This question of what distinguishes ecological or biological regularities from physical or chemical regularities is one of determining and evaluating exceptions and limits, the single and the individual. What distinguishes the law of gravity from a law governing the cycle of organic substances, the basic law of biocoenosis or from Mendel's laws? Most discussions revolve around the role of contingency, which often (but not always) cannot be eliminated in the biosciences. This is why, as John Beatty believes, there can be no laws in biology (on account of the contingency thesis) (J. Beatty, 'The Evolutionary Contingency Thesis', in G. Wolters, J. G. Lennox and P. McLaughlin (eds), *Concepts, Theories, and Rationality in Biological Sciences* (Pittsburgh, PA: University of Pittsburgh Press, 1995), pp. 45–81). And he adds that wherever it is possible to identify laws in biology, these are not 'biological' laws but ultimately physical or chemical laws. This has been countered by the argument that there certainly can be laws in biology, or at least law-like structures. Martin Carrier, for instance, argues that the extent to which laws underlie facts is more a matter of degree than of principle. The crucial point, he says, is to tolerate the idea that biological or ecological laws (the Lotka-Volterra case, for instance) apply to a variety of physically distinct cases. Carrier maintains that 'biology and the physical sciences are in the same ballpark. They both contain laws' (M. Carrier, 'Evolutionary Change and Lawlikeness: Beatty on Biological Generalizations', in G. Wolters and J. G. Lennox (eds), *Concepts, Theories, and Rationality in the Biological Sciences* (Konstanz/Pittsburgh, PA: Universitätsverlag Konstanz/ University of Pittsburgh Press, 1995), pp. 82–97, on pp. 92, 97).

19. G. J. Cooper, 'Theoretical Modeling and Biological Laws', *Philosophy of Science*, 63 (1996), pp. 28–35, on p. 33–4.

20. Waters, 'How Practical Know-how Contextualizes Theoretical Knowledge'.

21. S. Mitchell, 'Dimensions of Scientific Law', *Philosophy of Science*, 67 (2000), pp. 242–65.

22. K. Shrader-Frechette and E. McCoy, 'Applied Ecology and the Logic of Case Studies', *Philosophy of Science*, 61 (1994), pp. 228–49.

23. Ibid., p. 243.

24. Historian of ecology Robert McIntosh wrote an article on 'Pluralism in Ecology' in the *Annual Review of Ecology and Systematics*, 18 (1987), pp. 321–41, and a few years later on 'Concept and Terminology of Homogeneity and Heterogeneity in Ecology', in J. Kolasa and S. T. A. Pickett (eds), *Ecological Heterogeneity* (New York: Springer, 1991), pp. 24–46.

25. The study of such biotopes is called 'ecology of phytotelmata' and 'cave ecology'.

26. To be sure, this division of labour always refers to some physical dimension, though not necessarily to the same 'systematics': for botanists, what counts is biological systematics, whereas for molecular biologists it is biochemical systematics (what kind of molecules, proteins or nucleic acids are present), where the model organism can be either an animal or a plant.

27. Bachelard, *La formation de l'esprit scientifique*.

28. B. Latour, *Der Berliner Schlüssel: Erkundungen eines Liebhabers der Wissenschaften* (Berlin: Akademie Verlag, 1996).

29. W.-H. Roth and G. M. Bowen, 'Digitizing Lizards: The Topology of "Vision" in Ecological Fieldwork', *Social Studies of Science*, 29 (1999), pp. 719–64.

30. A. Shavit and J. Griesemer, 'There and Back Again, or the Problem of Locality in Biodiversity Surveys', *Philosophy of Science*, 76 (2009), pp. 273–94, on p. 273.

31. For further information, see the website for this project at www.mnh.si.edu/rc/fieldbooks/index.html; the collection of field books is accessible at collections.si.edu.

32. M R. Canfield (ed.), *Field Notes on Science and Nature* (Cambridge, MA: Harvard University Press, 2011).

33. Ibid., p. 16.

34. R. Kitching, 'A Reflection of the Truth', in M. Canfield (ed.), *Field Notes on Science and Nature* (Cambridge, MA: Harvard University Press, 2011), pp. 67–87, on p. 86.

35. See the more detailed discussion on this in Chapter 6 'Exploring the Field Ideal', the section headed 'Dirty Places of Experimentation'. Four key elements were identified, namely, the appreciation of a physical place, its institutional context, its specific practices and epistemological boundary work.

36. H. Arendt, *Vita activa*, 8th edn (Munich: Piper, 1994), p. 288.

37. This paints the picture of a world that evokes Bataille's general economy, in which a fully realized biosphere is no longer in the growth stage (in terms of biomass) but has become a 'luxurious squandering of energy in every form' (G. Bataille, *The Accursed Share: An Essay on General Economy*, trans. R. Hurley, 3 vols (New York: Zone Books, 1991), vol. 1, p. 33). This is discussed in detail in Chapter 13 'Political Economy of Experiments'.

38. Arendt, *Vita activa*, p. 115.

39. Ibid.

40. Goodin, *On Settling*, p. 17.

41. T. P. Hughes, *Human-built World: How to Think about Technology and Culture* (Chicago, IL: University of Chicago Press, 2004), p. 153.

42. Sarah Whatmore has given a nice theoretical account on green hybridities (S. Whatmore, *Hybrid Geographies: Natures, Cultures, Spaces* (London: Sage, 2002); a particular focus on climate change as a cultural phenomenon has been offered by H. Welzer and H.-G. Soeffner (eds), *KlimaKulturen: Soziale Wirklichkeiten im Klimawandel* (Frankfurt am Main: Campus, 2010).

43. 'Sustainable Development by Building with Nature", at http://www.ronaldwaterman.com/page10/page10.html.

44. There are now innumerable 'greenbelt' projects. Many cities, even smaller ones, have their one project of making running water visible or of linking together parks via greened paths. A nice case study on this widespread phenomenon is provided by Jens Lachmund for the city Berlin, starting in the period after reunification, when huge areas of urban wilderness (once prohibited zones) suddenly became available (J. Lachmund, *Greening Berlin: The Co-Production of Science, Politics, and Nature* (Cambridge, MA: MIT Press, 2012)). The phenomenon is also backed up by empirical studies including, for instance, the OPENspace Research Centre at Edinburgh, and particularly the study group around Catherine Ward Thompson investigating the relationship between green spaces and health benefits. 'New natures' are another phenomenon that is part of the greening culture, referring to the construction of functioning ecosystems in formerly industrialized regions (the construction of Lake Phoenix in the Ruhr city of Dortmund in Germany, for instance).

45. Ludwig Trepl (*Geschichte der Ökologie*) and, in particular, Gerhard Hard (*Spuren und Spurenleser: zur Theorie und Ästhetik des Spurenlesens in der Vegetation und anderswo* (Osnabrück: Universitätsverlag Rasch, 1995)) have pointed to the widespread tendency to 'naturalize' landscapes that owe their existence to formerly industrially or agriculturally exploited landscapes and that acquire and maintain their shape and appearance only through careful landscape management – in the case of the Lüneburger Heide, through sheep grazing. Other examples of such historical landscapes are the pre-industrial landscape, the peasant Alpine farm and the heath (Hard, *Spuren und Spurenleser*).

46. For the invention of the famous term, see W. Benjamin, *Das Kunstwerk im Zeitalter seiner technischen Reproduzierbarkeit* (Frankfurt am Main: Suhrkamp, 1969); a critical discussion is offered in Böhme, *Technik, Gesellschaft, Natur*.

47. On 'cabin ecology' specifically, see especially D. H. Calloway, *Human Ecology in Space Flight: Proceedings of the First International Interdisciplinary Conference* (New York: New York Academy of Sciences, Interdisciplinary Communications Program, 1965); and D. H. Calloway, *Human Ecology in Space Flight II: Proceedings of the Second International Interdisciplinary Conference* (New York: New York Academy of Sciences, Interdisciplinary Communications Program, 1967). On the subject of 'space biology', see for example J. S. Hanrahan and D. Bushnell, *Space Biology: The Human Factors in Space Flight* (New York: Thames & Hudson, 1960), as well as a host of magazine articles in, among others, *Missiles and Rockets, Astronautics, American Biology Teacher* and the *British Interplanetary Society Journal*.

48. See for instance A. C. Clarke, *The Exploration of Space* (New York: Harper, 1951). The programme still has strong technological as well as imaginary potential. It has played a role in recent space experiments as well as in trend-setting 'eco-design' prototypes. A good example of the former is the ongoing research project to develop 'aquatic modules for biogenerative life support systems' (V. Blüm, 'Aquatic Modules for Biogenerative Life Support Systems: Developmental Aspects Based on the Space Flight Results of the C.E.B.A.S. Mini-module', *Advances in Space Research*, 31 (2003), pp. 1683–91). For an eco-design product, see the air purifier 'Bel-Air' (2007), developed by Matthieu Lehanneur and David Edwards (Harvard University and Le Laboratoire Paris). It is based on a technology that was originally developed by NASA to improve air quality on board space shuttles (presented and discussed in S. Barbera and B. Cozzo, *Ecodesign* (Königswinter: Tandem-Verlag, 2009), p. 56).

49. R. Margalef, *Perspectives in Ecological Theory* (Chicago, IL: University of Chicago Press, 1968), p. 1.

50. K. E. Boulding, 'The Economics of the Coming Spaceship Earth', in H. Jarrett (ed.), *Environmental Quality in a Growing Economy* (Baltimore, MD: Johns Hopkins Univeristy Press, 1966), p. 3. See also S. Höhler and F. Luks (eds), *Beam us Up, Boulding! 40 Jahre 'Raumschiff Erde': Themenheft zum 40. Jubiläum von Kenneth E. Bouldings 'Operating Manual for Spaceship Earth' (1966)* (Hamburg: Vereinigung für Ökologische Ökonomie, 2006).

51. J. Lovelock, 'The Gaia Hypothesis', in P. Bunyard (ed.), *Gaia in Action: Science of the Living Earth* (Edinburgh: Floris, 1996), pp. 15–33.

52. See for instance the Preface of W. B. Cassidy (ed.), *Symposium on Bioengineering and Cabin Ecology* (Tarzana, AZ: American Association for the Advancement of Science, 1969); see also Peder Anker's eminent historical study on 'The Ecological Colonization of Space', *Environmental History*, 10 (2005), at http://www.historycooperative.org/journals/eh/10.2/anker.html [accessed 7 January 2009].

53. See for instance discussion of the 'Bios-3' facility at http://www.permanent.com/s-bios3.htm.

54. Mars project of the European Space Agency (ESA) in collaboration with the Russian Space Agency, at http://space.newscientist.com/article/dn11529 [accessed 16 September 2011]. The project was launched in 2010 and is a joint effort by the European and Russian space agencies. The experimental design is to house a group of people in a cabin for about 520 days – safely buried beneath the Earth's surface.

55. This has been discussed in more detail in A. Schwarz, 'Escaping from Limits into Visions of Space', in A. Ferrari and S. Gammel (eds), *Visionen der Nanotechnologie* (Berlin: Akademische Verlagsgesellschaft, 2009), a paper on which some passages in this chapter draw.

56. M. C. Roco and W. S. Bainbridge, *Societal Implications of Nanoscience and Nanotechnology*, NSET Workshop Report (Arlington, VA: NSET, 2001), p. 40.

57. 'The U.S. government needs a strategy for encouraging and stimulating green nanotechnology', D. Rejeski, Director of the Project on Emerging Technologies at the WWC (K. F. Schmidt, *Green Nanotechnology: It's Easier than You Think* (Washington DC: Woodrow Wilson International Center for Scholars, 2007), p. 3). Later in this report Rejeski states: 'It's not easy bein' green, but we think the United States is on track to be a global leader in green nanotech, and that the country's research and development portfolio and policy incentives should be directed toward this goal' (ibid., p. 23).

58. Barbara Karn, project leader at the US environmental agency, in a workshop-report entitled 'Nanotechnology and Life Cycle Assessment' (B. Karn and P. Aguar (organizers), 'Nanotechnology and Life Cycle Assessment', workshop report by W. Klöpffer (writing team cooridator) (Washington, DC: Woodrow Wilson International Center for Scholars, 2007).

11 On the Pleasures of Ecotechnology

1. M. Yourcenar, *Fires*, trans. D. Katz (Chicago, IL: University of Chicago Press 1994), p. 3.

2. Shellenberger and Nordhaus prompted fierce debate with their self-published essay 'The Death of Environmentalism' which provided the basis for *Break Through*. One of its most disputed claims is that environmentalism cannot deal with global warming because the issue is more complex than pollution problems. Also, they argue, American values have changed since the environmental movement's successes in the 1960s; thus it would be better if environmentalism faded away so that a new politics can be born in America (T. Nordhaus and M. Shellenberger, *Break Through: From the Death of Environmentalism to the Politics of Possibility* (Boston, MA: Houghton Mifflin Company, 2006)).

3. Comment by Richard Florida (author of *Rise of the Creative Class*); for more such statements of praise, see www.thebreakthrough.org [accessed 15 July 2011].

4. A. Rip, 'Folk Theories of Nanotechnologists', *Science as Culture*, 15 (2006), pp. 349–65, on p. 349.

5. This paragraph draws in part on A. Nordmann and A. Schwarz, 'The Lure of the "Yes": The Seductive Power of Technoscience', in M. Kaiser, M. Kurath, S. Maasen and C. Rehmann-Sutter (eds), *Assessment Regimes of Technology: Regulation, Deliberation and Identity Politics of Nanotechnology* (Dordrecht: Springer, 2010), pp. 255–77.

6. From 'Responsible Nanotechnology' website at http://crnano.typepad.com/crn-blog/2004/02/green_nanotechn.html [accessed 3 January 2008]; 'The Project on Emerging Nanotecologies' website at http://www.nanotechproject.org/publications/

archive/green_nanotechnology_its_easier_than [accessed 3 January 2008]; 'Phys.Org' website at http://www.physorg.com/news96781160.html [accessed 3 January 2008].

7. A report *Nanotechnologies for Sustainable Energy: Reducing Carbon Emissions through Clean Technologies and Renewable Energy Sources* (June 2007, at http://www.research-andmarkets.com/reports/470557) claims that current applications of nanotechnologies will result in a global annual saving of 8,000 tons of carbon dioxide in 2007, increasing to over a million tons by 2014. It also states that over the next seven years the biggest growth opportunities will come from the application of nanomaterials to the improved utilization of existing resources, rather than from new renewable energy technologies (the report is merchandised by 'Research and Markets').

8. This is not the place to elucidate the alternative moral economies that might support the game of seduction. If hedonism and the pursuit of pleasure appear rather obvious, this is only a sign of the times, which are characterized also by a remarkable optimism about nanotechnology's ability to cure the world's ills.

9. A logic of seduction was offered by M. Perniola, 'Logique de la seduction', *Traverses*, 18 (1980), pp. 2–9; this chapter mainly refers to the work of J. Baudrillard, *Seduction* (Houndsmills: Macmillan, 1990).

10. In light of standard conceptions that link power to interest, this sounds paradoxical. Power is believed to be a rationalizing expression, a rationalization of pure interest, and its effect the disenchantment of the world (K. Röttgers, *Spuren der Macht: Begriffsge-schichte und Systematik* (Freiburg im Breisgau: Alber, 1990)). And yet this 'magical' form of disinterested power is familiar from various accounts of the 'invisible hand'.

12 Conducting a Social Experiment: 'Building with Nature'

1. G. Valkenburg, 'Sustainable Technological Citizenship', *European Journal of Social Theory*, 15 (2012), pp. 471–87, on p. 473.

2. In their book *Social Natural Science* (*soziale Naturwissenschaft*) Böhme and Schramm introduced the term 'piece of nature' (*Naturstück*) (G. Böhme and E. Schramm (eds), *Soziale Naturwissenschaft: Wege zu einer Erweiterung der Ökologie* (Frankfurt am Main: Fischer Taschenbuch Verlag, 1984)) which was subsequently adopted by several authors in the domain of environmental philosophy, e.g. by Werner Ingensiep, who published a book on 'NaturStücke' (H. W. Ingensiep and R. Hoppe-Sailer, *NaturStücke: Zur Kulturgeschichte der Natur* (Ostfildern: edition tertium, 1996)).

3. Some remarks relating to the economic dimensions of the project: the total costs of realizing the project amounted to EUR 380 million, 30 million of which were spent on so-called compensation measures, the most prominent being the 'near-natural by-pass channel'. The average production per year of the power plant is 600 million kilowatts. The operator Energiedienst sells electricity to the nearby carbon (Evonik) and aluminium industry in Germany as well as to municipalities in Switzerland and Germany.

4. Helmut Reif (chief engineer for the overall project): 'While working with the commission, an extremely high level of trust built up over time'. Reif also pointed out that it was very easy to include the stakeholders in the discussions and that no ideological obstacles arose. Rolf-Jürgen Gebler (responsible for the design, planning and construction of the by-pass): 'I experienced the collaboration (with the monitoring commission) as being very beneficial. During the construction phase we had to deal with modifications that could be discussed immediately and frankly with the commission. This enabled diffi-

culties to be resolved very quickly. I've only experienced this kind of committee with projects in the border zone with Switzerland'.

5. For a more detailed discussion of this motif, see the section on 'Restoration Ecology' in Chapter 10.

6. Of course, there are also 2D representations that were used, the already mentioned topographical and hydrological maps and mainly the plan delivered by the landscape designer and engineer Rolf Gebler, which is developed in detail elsewhere.

7. Along with suitability for a wide spectrum of fauna, habitat enrichment for rheophilic species, flexible instead of static construction, toleration of limited erosion and aggradation.

8. For example: transgressive nature conservation, design as a medium of political deliberation, invasive models, landscape as a simulacrum.

13 Political Economy of Experiments

1. W. Herrndorf, *Arbeit und Struktur* (Berlin: Rowohlt, 2013), p. 56.

2. On the notion of a moral economy of science, see Daston, 'The Moral Economy of Science'. Ernst Mach argued that concepts serve to economize the multiplicity of sensations (E. Mach, *Analysis of Sensations* (New York: Dover, 1959)).

3. H. Lacey, 'The Life and Times of Transgenics', in B. Bensaude-Vincent, S. Loeve, A. Nordmann and A. Schwarz (eds), *Attractive Objects* (forthcoming).

4. The choice between adaptation to and the conquest of limits is politically salient especially in current debates about the proper response to global warming, where adaptationist proposals are countered by the hope that new technologies (including geo-engineering) can sustain further economic growth.

5. Some parts of this chapter are taken from an already published paper (Schwarz and Nordmann, 'Unlimited Potential in a Limited World').

6. For instance in B. Woods, 'Political Economy of Science', in G. Ritzer (ed.), *The Blackwell Encyclopedia of Sociology* (Oxford: Blackwell Publishing, 2007), pp. 3436–9; or also in B. R. Martin and P. Nightingale (eds), *The Political Economy of Science, Technology and Innovation* (Cheltenham: Edward Elgar Publishing, 2000); but also H. Rose and S. Rose, *The Political Economy of Science: Ideology of/in the Natural Sciences* (London: Macmillan, 1976).

7. Bataille, *The Accursed Share*, p. 29. Bataille uses references sparingly, but here ('*'), Bataille refers directly to the author who first conceptualized the term: '*See Vernadsky 1929, where some of the considerations that follow are outlined (from a different viewpoint)'.

8. Ibid., p. 30 (translation modified by Schwarz/Nordmann).

9. Ibid., p. 33

10. Ibid., p. 64. This brings to mind Hannah Arendt's *Homo laborans* (see the section 'Ecotechnology: On Connections in Space' in Chapter 10).

11. See also the discussion on the different roles of scientific laws in the natural sciences in '"Ecology" as a Snapshot in the History of Human-Environment Relationships' in Chapter 10.

12. In ongoing debates about the limits of global resources, scientists have identified a 'new scarcity' in resource use. They focus especially on 'the big three': land use change (from cropland to industrial/urban land), emission of greenhouse gases and extraction of materials (S. Bringezu, *Sustainable Resource Management: Global Trends, Visions and Policies*

(Sheffield: Greenleaf Publishing, 2009)). These 'big three' are presented as a technological challenge rather than as a requirement to adapt.

13. A. Lavoisier, *Traité élémentaire de chimie* (Chicago, IL: Encyclopaedia Britannica, 1952), p. 41.

14. B. Bensaude-Vincent, 'The Balance: Between Chemistry and Politics', *Eighteenth Century*, 33 (1992), pp. 217–37; M. Carrier, 'Antoine Laurent de Lavoisier und die Chemische Revolution', in A. Schwarz and A. Nordmann (eds), *Das bunte Gewand der Theorie* (Freiburg: Alber, 2009), pp. 12–42.

15. In her book on *A History of the Scientific Fact*, Mary Poovey offers a knowledgeable study, rich in historical details and presenting a sophisticated alignment of philosophical arguments.

16. Bataille's caricature of restricted economics agrees with classical economics, especially insofar as it aims to become properly scientific by producing general testable models of economic exchange. Thus the characterization does not do justice to the current state of economics as a science and technoscience.

17. Bataille, *The Accursed Share*, pp. 22–3. Again, the wording of the translation is slightly altered.

18. Ibid., pp. 10–11.

19. G. Bachelard, G., *The Philosophy of No: A Philosophy of the Scientific Mind* (New York: Orion Press, 1968).

20. Bensaude-Vincent, 'The Balance'; Holmes, *Eighteenth-Century Chemistry as an Investigative Enterprise*.

21. H. Arendt, *The Human Condition*, 2nd edn (Chicago, IL: University of Chicago Press, 1998), p. 1.

22. Bacon's tools for transgressing limits were facts and aphorisms: facts are the driving force in experimentation, while aphorisms invoke possibilities. This was discussed in the section on 'Fragments in Aphorisms; An Active Form of Learning' in Chapter 1.

23. C. Becker, M. Faber, K. Hertel and R. Manstetten, 'Die unterschiedlichen Sichtweisen von Malthus und Wordsworth auf Mensch, Natur und Wirtschaft', in F. Beckenbach, U. Hampicke and C. Leipert (eds), *Soziale Nachhaltigkeit: Jahrbuch Ökologische Ökonomik*, Vol. 5 (Marburg: Metropolis, 2007), pp. 275–99.

24. T. R. Malthus, *An Essay on the Principle of Population, as it Affects the Future Improvement of Society* (London: Macmillan & Co., 1798), pp. 13–15.

25. Ibid., pp. 127–8.

26. Ibid., p. 207.

27. This is not the place to present the argument in detail. Georgescu-Roegen is one of the best-known scholars who relied on the work of systems biologist Bertalanffy. Bertalanffy, in turn, began his career thinking about systems biology in the 1930s by adapting the two laws of thermodynamics to biology and transforming them into principles of gestalt.

28. R. L. Lindeman, 'The Trophic-dynamic Aspect of Ecology', *Ecology*, 23 (1942), pp. 399–418, on p. 400.

29. The concept had already been invented by geographer Eduard Suess, but it was only Vernadsky who conceptualized the biosphere as it was taken up by Bataille and as we know it today.

30. V. I. Vernadsky, *La Biosphère* (Paris: Alcan, 1929).

31. The Macy Conferences (with participants such as Norbert Wiener, John von Neumann, Warren McCulloch, Margaret Mead and Heinz von Foerster) contributed decisively towards the dissemination of cybernetic approaches beyond primarily technological

applications into areas such as the social sciences, psychology, ecology and the human and life sciences in general. For more detail see C. Pias (ed.), *Cybernetics – Kybernetik: The Macy-Conferences (1946–1953)*, 2 vols (Zurich: Diaphanes, 2003).

32. The formation of the GST is described in W. Gray and N. D. Rizzo (eds), *Unity through Diversity: Festschrift for Ludwig von Bertalanffy* (New York, London and Paris: Gordon and Breach Science Publishers, 1973).

33. Astrid Schwarz offers a closer look at the beginnings of systems theory in biology and ecology with a special emphasis on the concept of gestalt (A. Schwarz, 'Gestalten werden Systeme: Frühe Systemtheorie in der Ökologie', in K. Mathes, B. Breckling and K. Ekschmidt (eds), *Systemtheorie in der Ökologie* (Landsberg: Ekomed, 1996), pp. 35–45). A detailed story of systems theory in early ecology is given in A. Voigt, 'The Rise of Systems Theory in Ecology', in A. Schwarz and K. Jax (eds), *Ecology Revisited: Reflecting on Concepts, Advancing Science* (Dordrecht: Springer, 2011), pp. 183–94.

34. G. E. Hutchinson, 'Circular Causal Systems in Ecology', *Annals of the New York Academy of Sciences*, 50 (1948), pp. 221–46. George Evelyn Hutchinson participated in a number of Macy Conferences.

35. For more details, see the Declaration of the World Resouces Forum at www.worldresourcesforum.org/wrf_declaration [accessed 15 January 2010]. The WRF was founded in Davos, Switzerland, in September 2009.

36. A vivid illustration of this was provided in a large exhibit curated by the German Max Planck Society for basic research. It presented as a point of departure a reminder of resource limits. From then on, however, it featured the power of the technosciences to go beyond these limits: 'we must grow beyond ourselves' (Max-Planck-Gesellschaft, *Expedition Zukunft: Science Express* (Munich: Max Planck Gesellschaft, 2009), pp. 181, 187).

37. UN Secretary General, *Progress to Date and Remaining Gaps in the Implementation of Outcomes of the Major Summits in the Area of Sustainable Development, as well as the Analysis of the Themes of the Conference*, Report for the Preparatory Committee for the United Nations Conference on Sustainable Development (New York: United Nations, 2010), p. 15.

38. With regard to the green economy, see among others the work of Ulrich Brand (e.g. 'Green Economy – the Next Oxymoron?', *Gaia*, 21 (2012), pp. 28–32); the green new deal has been commented on critically in a study by Frank Adler and Ulrich Schachtschneider (F. Adler and U. Schachtschneider, *Green New Deal, Suffizienz oder Ökosozialismus? Konzepte für gesellschaftliche Wege aus der Ökokrise* (Munich: Oekom, 2010)); the importance and at once underestimation of maintenance and sufficiency has been put forward in a compelling study by David Edgerton called *The Shock of the Old* (London: Profile Books, 2008).

WORKS CITED

Achinstein, P., 'The War on Induction: Whewell Takes on Newton and Mill (Norton Takes on Everyone)', *Philosophy of Science*, 77 (2010), pp. 728–39.

Adler, F., and U. Schachtschneider, *Green New Deal, Suffizienz oder Ökosozialismus? Konzepte für gesellschaftliche Wege aus der Ökokrise* (Munich: Oekom, 2010).

Allenby, B. R., and D. Rejeski, 'The Industrial Ecology of Emerging Technologies', *Journal of Industrial Ecology*, 12 (2008), pp. 267–9.

Amann, K., 'Menschen, Mäuse und Fliegen: Eine wissenssoziologische Analyse der Transformation von Organismen in epistemische Objekte', *Zeitschrift für Soziologie*, 23 (1994), pp. 22–40.

Anders, G., *Die Antiquiertheit des Menschen: Über die Seele im Zeitalter der zweiten industriellen Revolution* (Munich: Beck, 1956).

Anker, P., 'The Ecological Colonization of Space', *Environmental History*, 10 (2005), at http://www.historycooperative.org/journals/eh/10.2/anker.html [accessed 7 July 2009].

Appiah, K. A., *Experiments in Ethics* (Cambridge, MA: Harvard University Press, 2008).

—, *Presidential Address*, Eastern Division APA (December 2007), at http://appiah.net/wp-content/uploads/2010/10/APA-Lecture-2007-for-Web.pdf [accessed 12 September 2013].

Arendt, H., 'Social Science Techniques and the Study of Concentration Camps', *Jewish Social Studies*, 12 (1950), pp. 49–64.

—, *Vita activa*, 8th edn (Munich: Piper, 1994).

—, *The Human Condition*, 2nd edn (Chicago, IL: University of Chicago Press, 1998).

Bachelard, G., *La formation de l'esprit scientifique: Contribution à une psychoanalyse de la connaissance objective* (Paris: Vrin, 1938).

—, *L'activité rationaliste de la physique contemporaine* (Paris: Presses universitaires de France, 1951).

—, *The Philosophy of No: A Philosophy of the Scientific Mind* (New York: Orion Press, 1968).

—, *Epistemologie: Ausgewählte Texte* (Frankfurt am Main: Ullstein, 1974).

Bacon, F., *The Novum Organum*, trans. M. D. B. D. (London: Thomas Lee at the Turkshead, 1676).

—, *Neues Organon: Vorrede: Erstes Buch*, ed. W. Krohn, 2 vols (Hamburg: Meiner, 1990).

—, *The Advancement of Learning*, ed. M. Kiernan, 15 vols (Oxford: Clarendon Press, 2000).

—, *The Instauratio Magna Part II: Novum Organum and Associated Texts*, ed. and trans. G. Rees and M. Wakely, 15 vols (Oxford: Clarendon Press, 2004).

Bainbridge, W., and M. Roco (eds), *Converging Technologies for Improving Human Performance: Nanotechnology, Biotechnology, Information Technology and Cognitive Science*, NSF/DOC-Sponsored Report (Arlington, VA, prepublication online version, 2002), at http:// http://www.markswatson.com/NBIC.pdf [accessed 17 September 2013].

Baird, D., 'Thing Knowledge', in H. Radder (ed.), *The Philosophy of Scientific Experimentation* (Pittsburgh, PA: University of Pittsburgh Press, 2003), pp. 39–67.

—, *Thing Knowledge: A Philosophy of Scientific Instruments* (Berkeley, CA: University of California Press, 2004).

Balsiger, P. W., 'Überlegungen und Bemerkungen hinsichtlich einer Methodologie interdisziplinärer Wissenschaftspraxis', in P. W. Balsiger, R. Defila and A. di Giulio (eds), *Ökologie und Interdisziplinarität: eine Beziehung mit Zukunft?* (Basel: Birkhäuser, 1996), pp. 73–85.

Barbera, S., and B. Cozzo, *Ecodesign* (Königswinter: Tandem-Verlag, 2009).

Barthes, R., *Mythologies* (New York: Hill and Wang, 1972).

Bataille, G., *La part maudite*, 12 vols (Paris: Gallimard, 1949).

—, *The Accursed Share: An Essay on General Economy*, trans. R. Hurley, 3 vols (New York: Zone Books, 1991).

Baudrillard, J., *Seduction* (Houndsmills: Macmillan, 1990).

Beatty, J., 'The Evolutionary Contingency Thesis', in G. Wolters, J. G. Lennox and P. McLaughlin (eds), *Concepts, Theories, and Rationality in Biological Sciences* (Pittsburgh, PA: University of Pittsburgh Press, 1995), pp. 45–81.

Beck, U., *Risikogesellschaft: Auf dem Weg in eine andere Moderne* (Frankfurt am Main: Suhrkamp, 1986).

—, 'Die blaue Blume der Moderne', *Der Spiegel*, 33 (1991), pp. 50–1.

—, and C. Lau, *Entgrenzung und Entscheidung* (Frankfurt am Main: Suhrkamp, 2004).

Becker, C., M. Faber, K. Hertel and R. Manstetten, 'Die unterschiedlichen Sichtweisen von Malthus und Wordsworth auf Mensch, Natur und Wirtschaft', in F. Beckenbach, U. Hampicke and C. Leipert (eds), *Soziale Nachhaltigkeit: Jahrbuch Ökologische Ökonomik*, Vol. 5 (Marburg: Metropolis, 2007), pp. 275–99.

Becker, E., 'Transformations of Social and Ecological Issues into Transdisciplinary Research', in UNESCO/EOLSS Publishers (ed.), *Knowledge for Sustainable Development: An Insight into the Encyclopedia of Life Support Systems*, 3 vols (Paris and Oxford: EOLSS Publishers, 2002), vol. 3, pp. 949–63.

Benjamin, W., *Das Kunstwerk im Zeitalter seiner technischen Reproduzierbarkeit* (Frankfurt am Main: Suhrkamp, 1969).

Bensaude-Vincent, B., 'The Balance: Between Chemistry and Politics', *Eighteenth Century*, 33 (1992), pp. 217–37.

—, *Dans le laboratoire de Lavoisier* (Paris: Nathan, Monde en Poche, 1993).

—, 'Chemistry in the French Tradition of Philosophy of Science: Duhem, Meyerson, Metzger and Bachelard', *Studies in History and Philosophy of Science*, 36 (2005), pp. 627–48.

—, S. Loeve, A. Nordmann and A. Schwarz, 'Matters of Interest: The Objects of Research in Science and Technoscience', *Journal for General Philosophy of Science*, 42 (2011), pp. 365–83.

Berkeley, G., *A Treatise Concerning the Principles of Human Knowledge – Part I: Wherein the Chief Causes of Error and Difficulty in the Sciences, with the Grounds of Scepticism, Atheism, and Irreligion, are Inquir'd* (Dublin: Jeremy Pepyat, 1710), reproduction of original from British Library, University of Basel online, at http://find.galegroup.com/ecco/infomark.do?&source=gale&prodId=ECCO&userGroupName=unibas&tabID=T001&docId=CW118263402&type=multipage&contentSet=ECCOArticles&version=1.0&docLevel=FAACSIMILE [accessed 3 March 2011].

Billick, I., and M. V. Price (eds), *The Ecology of Place: Contributions of Place-based Research to Ecological Understanding* (Chicago, IL: University of Chicago Press, 2010).

Bliss, P. (ed.), *The Heart of Emerson's Journals* (New York: Dover Publications, 1995).

Blüm, V., 'Aquatic Modules for Biogenerative Life Support Systems: Developmental Aspects Based on the Space Flight Results of the C.E.B.A.S. Mini-module', *Advances in Space Research*, 31 (2003), pp. 1683–91.

Böhme, G., 'Alternatives in Science – Alternatives to Science?', in H. Nowotny and H. Rose (eds), *Counter-Movements in the Sciences: The Sociology of the Alternatives to Big Science* (Dordrecht: D. Reidel, 1979), pp. 105–25.

—, *Technik, Gesellschaft, Natur: Die zweite Dekade des interdisziplinären Kolloquiums der Technischen Hochschule Darmstadt* (Darmstadt: Technische Hochschule Darmstadt, 1992).

—, and G. Schiemann, *Phänomenologie der Natur* (Frankfurt am Main: Suhrkamp, 1997).

—, and E. Schramm (eds), *Soziale Naturwissenschaft: Wege zu einer Erweiterung der Ökologie* (Frankfurt am Main: Fischer Taschenbuch Verlag, 1984).

—, and N. Stehr (eds), *The Knowledge Society: The Growing Impact of Scientific Knowledge on Social Relations* (Dordrecht: D. Reidel, 1986).

—, W. van der Daele and W. Krohn, *Experimentelle Philosophie: Ursprünge autonomer Wissenschaftsentwicklung* (Frankfurt am Main: Suhrkamp, 1977).

Boulding, K. E., 'The Economics of the Coming Spaceship Earth', in H. Jarrett (ed.), *Environmental Quality in a Growing Economy* (Baltimore, MD: Johns Hopkins University Press, 1966), pp. 3–14.

Bowen, M., *Empiricism and Geographical Thought: From Francis Bacon to Alexander von Humboldt* (Cambridge: Cambridge University Press, 1981).

Brand, U., 'Green Economy – the Next Oxymoron?', *Gaia*, 21 (2012), pp. 28–32.

Bringezu, S., *Sustainable Resource Management: Global Trends, Visions and Policies* (Sheffield: Greenleaf Publishing, 2009).

Brönnimann, S., and G. H. Hadorn, 'Lessons for Science from the "Year without a Summer" of 1816: What does it Take for Science to Respond to Climatic Change?', *Gaia*, 22 (2013), pp. 169–73.

Brothwell, D. R. (ed.), *Environmental and Experimental Studies in the History of Archaeology*, Symposia of the Association for Environmental Archaeology No. 9, Roskilde, Denmark, 1988 (Oxford: Oxbow Books, 1990).

Bundesministerium für Umwelt und Reaktorsicherheit (BMU) (ed.), *Verantwortlicher Umgang mit Nanotechnologien. Bericht und Empfehlungen der NanoKommission 2011*, text written by A. Grobe, ed. W.-M. Catenhusen (Berlin: BMU, 2010).

Burian, R. M., 'Comments on Complexity and Experimentation in Biology', *Philosophy of Science*, 64 (1997) (Supplement, Proceedings of the 1996 Biennial Meetings of the Philosophy of Science Association, Part II: Symposia Papers), pp. 279–91.

Burkamp, W., *Die Struktur der Ganzheiten* (Berlin: Junker und Dünnhaup, 1929).

Butts, R. E., 'Introduction', in W. Whewell, *Theory of Scientific Method*, ed. R. E. Butts (Indianapolis, IN: Hackett Publishing, 1989), pp. 3–30.

Callon, M., 'Four Models for the Dynamics of Science', in S. Jasanoff, G. E. Markle, J. C. Peterson and Trevor Pinch (eds), *Handbook of Science and Technology Studies* (Thousand Oaks, CA: Sage, 1995), pp. 29–63.

Calloway, D. H., *Human Ecology in Space Flight: Proceedings of the First International Interdisciplinary Conference* (New York: New York Academy of Sciences, Interdisciplinary Communications Program, 1965).

—, *Human Ecology in Space Flight II: Proceedings of the Second International Interdisciplinary Conference* (New York: New York Academy of Sciences, Interdisciplinary Communications Program, 1967).

Campbell, D. T., 'Reforms as Experiments', *American Psychologist*, 24 (1969), pp. 409–29.

—, and J. C. Stanley, 'Experimental and Quasi-experimental Designs for Research on Teaching', in N. L. Gage (ed.), *Handbook of Research and Teaching* (Chicago, IL: Rand McNally, 1963), pp. 171–246.

Canfield, M. R., *Field Notes on Science and Nature* (Cambridge, MA: Harvard University Press, 2011).

Canguilhem, G., *Wissenschaftsgeschichte und Epistemologie: Gesammelte Aufsätze*, trans. M. Bischof and W. Seutter, ed. W. Lepenies (Frankfurt am Main: Suhrkamp, 1979).

Caquet, T., L. Lagadic and S. R. Sheffield, 'Mesocosms in Ecotoxicology. Outdoor Aquatic Systems', *Reviews of Environmental Contamination and Toxicology*, 165 (2000), pp. 1–38.

Carnap, R., 'Meaning and Necessity: A Study in Semantics and Modal Logic', *Revue Internationale de Philosophie*, 4 (1950; reprinted by University of Chicago Press, 1956), pp. 20–40.

Carr, B., 'Popper's Third World', *Philosophical Quarterly*, 27 (1977), p. 219.

Carrier, M., 'Evolutionary Change and Lawlikeness: Beatty on Biological Generalizations', in G. Wolters and J. G. Lennox (eds), *Concepts, Theories, and Rationality in the Biological Sciences* (Konstanz/Pittsburgh, PA: Universitätsverlag Konstanz/University of Pittsburgh Press, 1995), pp. 82–97.

—, *Wissenschaftstheorie: Zur Einführung* (Hamburg: Junius, 2006).

—, 'Theories for Use: On the Bearing of Basic Science on Practical Problems', in *EPSA07: 1st Conference of the European Philosophy of Science Association* (Madrid, 15–17 November 2007), p. 32, at http://philsci-archive.pitt.edu/id/eprint/3690 [accessed 21 December 2013].

—, 'Antoine Laurent de Lavoisier und die Chemische Revolution', in A. Schwarz and A. Nordmann (eds), *Das bunte Gewand der Theorie* (Freiburg: Alber, 2009), pp. 12–42.

Cartwright, N., *The Dappled World: A Study of the Boundaries of Science* (Cambridge: Cambridge University Press, 1999).

Cassidy, W. B. (ed.), *Symposium on Bioengineering and Cabin Ecology* (Tarzana, AZ: American Association for the Advancement of Science, 1969).

Cassirer, E., *Erkenntnis, Begriff, Kultur*, ed. and commentary by R. A. Bast (Hamburg: Felix Meiner, 1993).

Cathcart, T., and D. Klein, *Plato and a Platypus Walk into a Bar: Understanding Philosophy through Jokes* (New York: Penguin Books, 2007).

Chakravartty, A., 'Semirealism', *Studies in History and Philosophy of Science*, 29 (1998), pp. 391–408.

Chalmers, A. F., *What is this Thing Called Science?* (1976; Queensland: University of Queensland Press, 1999).

Christie, J. R. R., 'Aurora, Nemesis and Clio', *British Journal for the History of Science*, 26 (1993), pp. 391–405.

Cittadino, E., *Nature as the Laboratory: Darwinian Plant Ecology in the German Empire, 1880–1900* (Cambridge: Cambridge University Press, 1990).

Clarke, A. C., *The Exploration of Space* (New York: Harper, 1951).

Clift, R., and S. Lloyd, 'Nanotechnology. A New Organism in the Industrial Ecosystem?', *Journal of Industrial Ecology*, 12 (2008), pp. 259–62.

Cooper, G. J., 'Theoretical Modeling and Biological Laws', *Philosophy of Science*, 63 (1996), pp. 28–35.

Cunningham, A., and P. Williams (eds), *The Laboratory Revolution in Medicine* (Cambridge: Cambridge University Press, 1992).

Daston, L., 'The Moral Economy of Science', *Osiris*, 10 (1995), pp. 3–24.

—, 'The Empire of Observation, 1600–1800', in L. Daston and E. Lunbeck (eds), *Histories of Scientific Observation* (Chicago, IL: University of Chicago Press, 2011), pp. 81–113.

—, and E. Lunbeck, 'Introduction', in L. Daston and E. Lunbeck (eds), *Histories of Scientific Observation* (Chicago, IL: University of Chicago Press, 2011), pp. 1–9.

Descola, P., and G. Palsson, *Nature and Society: Anthropological Perspectives* (London: Routledge, 1996).

Dewey, J., *Context and Thought* (Berkeley, CA: University of California Publications in Philosophy, 1929).

—, *Erfahrung und Natur*, trans. M. Suhr (Frankfurt am Main: Suhrkamp, 2007).

Diderot, M., and M. d'Alembert (eds), *Encyclopédie, ou, Dictionnaire raisonné des sciences, des arts et des métiers* (Lausanne and Berne: Sociétés typographiques, 1779–82).

Dierig, S., 'Engines for Experiment: Laboratory Revolution and Industrial Labor in the Nineteenth-Century City', *Osiris*, 18 (2003), pp. 116–34.

Dietz, H., and T. Steinlein, 'Recent Advances in Understanding Plant Invasions', *Progress in Botany*, 65 (2004), pp. 539–73.

Donaldson, S. I., 'In Search of the Blueprint for an Evidence-based Global Society', in S. I. Donaldson, C. A. Christie and M. M. Mark (eds), *What Counts as Credible Evidence in Applied Research and Contemporary Evaluation Practice?* (Newbury Park, CA: Sage, 2008), pp. 2–18.

Duncan, J., 'The Hidden Geographies of Social Sciences and the Myth of the Geographical Turn', *Environment and Planning D: Society and Space*, 13 (1995), pp. 379–80.

Ede, A., and L. B. Cormack, *A History of Science in Society: From Philosophy to Utility* (Peterborough: Broadview Press, 2004).

Edgerton, D., *The Shock of the Old* (London: Profile Books, 2008).

Einsele, W., 'Die Umsetzung von zugeführtem, anorganischen Phosphat im eutrophen See und ihre Rückwirkung auf seinen Gesamthaushalt', *Zeitschrift für Fischerei und deren Hilfswissenschaften*, 39 (1941), pp. 407–88.

Eisel, U., 'Triumph des Lebens: Der Sieg christlicher Wissenschaft über den Tod in Arkadien', *Urbs et Regio Sonderband*, 65 (1997), pp. 39–160.

Elzinga, A., 'The New Production of Particularism in Models Relating to Research Policy: A Critique of Mode 2 and Triple Helix', *Contribution to the 4S-EASST Conference* (Paris, 2004), at www.csi.ensmp.fr/WebCSI/4S/download_paper/download_paper.php?paper=elzinga.pdf [accessed 22 December 2013].

Etzkowitz, H., 'Innovation in Innovation: The Triple Helix of University-Industry-Government Relations', *Social Science Information*, 42 (2003), pp. 293–337.

Farrington, B., *Francis Bacon: Philosopher of Industrial Science.* (London: Lawrence and Wishart, 1951).

Felt, U. (rapporteur), B. Wynne (chairman), M. Callon, M. E. Gonçalves, S. Jasanoff, M. Jepsen, B. Joly, Z. Konopasek, S. May, C. Neubauer, A. Rip, K. Siune, A. Stirling and M. Tallacchini, *Taking European Knowledge Society Seriously: Report of the Expert Group on Science and Governance to the Science, Economy and Society Directorate* (Brussels: Directorate-General for Research, European Commission, 2007).

Ferrari, A., *Genmaus & Co.: Gentechnisch veränderte Tiere in der Biomedizin* (Erlangen: Harald Fischer Verlag, 2008).

Fisch, M., 'Antithetical Knowledge', in M. Fisch and S. Schaffer (eds), *William Whewell: A Composite Portrait* (Oxford: Clarendon Press, 1991), pp. 289–310.

—, and S. Schaffer, *William Whewell: A Composite Portrait* (Oxford: Clarendon Press, 1991).

Fischer, M., and M. A. Hajer (eds), *Living with Nature: Environmental Politics as Cultural Discourse* (New York: Oxford University Press, 1999).

Fleck, L., *Genesis and Development of a Scientific Fact*, trans. F. Bradley and T. J. Trenn (Chicago, IL: University of Chicago Press, 1979).

—, *Erfahrung und Tatsache*, ed. L. Schäfer and T. Schnelle (Frankfurt am Main: Suhrkamp, 1983).

—, *Denkstile und Tatsachen* (Berlin: Suhrkamp, 2011).

Forman, P., 'The Primacy of Science in Modernity, of Technology in Postmodernity, and of Ideology in the History of Technology', *History and Technology*, 23 (2007), pp. 1–152.

Foucault, M., *Les mots et les choses* (Paris: Gallimard, 1966).

—, *The Order of Things: An Archaeology of the Human Sciences* (New York: Pantheon Books, 1970).

—, 'Of Other Spaces', *Diacritics*, 16 (1986), pp. 22–7.

Frodeman, R., *Geo-logic: Breaking Ground between Philosophy and the Earth Sciences* (Albany, NY: State University of New York Press, 2003).

Funtowicz, S., and J. Ravetz, 'Uncertainty, Complexity and Post-normal Science', *Environmental Toxicology and Chemistry*, 13 (1994), pp. 1881–5.

Galavotti, M. C. (ed.), *Observation and Experiment in the Natural and Social Sciences* (Dordrecht: Kluwer Academic Publishers 2003).

Galison, P., *How Experiments End* (Chicago, IL: University of Chicago Press, 1987).

—, 'History, Philosophy, and the Central Metaphor', *Science in Context*, 2 (1988), pp. 197–212.

—, *Image and Logic: A Material Culture of Microphysics* (Chicago, IL: University of Chicago Press, 1997).

—, and D. J. Stump (eds), *The Disunity of Science: Boundaries, Contexts, and Power* (Stanford, CA: Stanford University Press, 1996).

—, and E. Thompson (eds), *The Architecture of Science* (Cambridge, MA: MIT Press, 1999).

Gaukroger, S., *Francis Bacon and the Transformation of Early-Modern Philosophy* (Cambridge: Cambridge University Press, 2001).

Georges, K. E., *Ausführliches Lateinisch-Deutsches Handwörterbuch*, 2 vols (Basel: Benno Schwabe, 1959).

Gibbons, M., C. Limoges, H. Nowotny, S. Schwartzmann, P. Scott and M. Trow, *The New Production of Knowledge: The Dynamics of Science and Research in Contemporary Societies* (London: Sage Publications, 1994).

Gibson, J. J., 'New Reasons for Realism', *Synthese*, 17 (1967), pp. 401–18.

—, *The Ecological Approach to Visual Perception* (Boston, MA: Houghton Mifflin, 1979).

Gieryn, T., 'Boundary Work and the Demarcation of Science from Non-science: Strains and Interests in Professional Ideologies of Scientists', *American Sociological Review*, 48 (1983), pp. 781–95.

—, 'Boundaries of Science', in S. Jasanoff (ed.), *Handbook of Science and Technology Studies* (London: Sage Publications 1995), pp. 393–443.

—, *Cultural Boundaries of Science: Credibility on the Line* (Chicago, IL: University of Chicago Press, 1999).

—, 'City as Truth-Spot: Laboratories and Field-Sites in Urban Studies', *Social Studies of Science*, 36 (2006), pp. 5–38.

Gill, S. B., and J. Borchers, 'Knowledge in Co-Action', in S. B. Gill (ed.), *Cognition, Communication, and Interaction* (Dordrecht: Springer, 2008), pp. 38–55.

Golan, T., *Laws of Men and Laws of Nature: The History of Scientific Expert Testimony in England and America* (Cambridge, MA: Harvard University Press, 2004).

Goliszek, A., *In the Name of Science: A History of Secret Programs, Medical Research, and Human Experimentation* (New York: St Martin's Press, 2003).

Goodin, R. E., *On Settling* (Princeton, NJ: Princeton University Press, 2012).

Gray, W., and N. D. Rizzo (eds), *Unity through Diversity: Festschrift for Ludwig von Berta-lanffy* (New York, London and Paris: Gordon and Breach Science Publishers, 1973).

Griesemer, J. R., 'Taking Organic Processes: Representations and Research Styles in Classical Embryology and Genetics', in M. D. Laubichler and J. Maienschein (eds), *From Embryol-ogy to Eco-devo* (Cambridge, MA: MIT Press, 2007), pp. 375–433.

Gross, M., and W. Krohn, 'Science in a Real World Context: Constructing Knowledge through Recursive Learning', *Philosophy Today*, 48 (2004), pp. 38–50.

—, H. Hoffmann-Riem and W. Krohn, *Realexperimente: Ökologische Gestaltungsprozesse in der Wissensgesellschaft* (Bielefeld: Transcript, 2005).

Guerrini, A., 'Animal Experiments and Antivivisection Debates in the 1820s', in C. Knellwolf and J. Goodall (eds), *Frankenstein's Science: Experimentation and Discovery in Romantic Culture, 1780–1830* (Aldershot and Burlington, VT: Ashgate, 2008), pp. 71–86.

Habermas, J., *Moral Consciousness and Communicative Action*, trans C. Lenhardt and S. W. Nicholsen (Cambridge: Polity Press, 1990).

—, *Der gespaltene Westen: Kleine politische Schriften* (Frankfurt am Main: Suhrkamp, 2004).

Hacking, I., *Representing and Intervening* (Cambridge: Cambridge University Press, 1983).

—, 'Another World is being Constructed Right Now: The Ultracold', in H.-J. Rheinberger (ed.), *Conference 'the Shape of Experiment'*, Berlin, 2–5 June 2005, Preprint 318 (Berlin: MPI für Wissenschaftsgeschichte, 2006), pp. 15–44.

Hadorn, G. H., U. Kunz and M. Maibach, 'Umweltforschung: Nice to Have or Need to Have?', *Gaia*, 13 (2004), pp. 70–3.

Hanrahan, J. S., and D. Bushnell, *Space Biology: The Human Factors in Space Flight* (New York: Thames & Hudson, 1960).

Hanson, E., *Discovering the Subject in Renaissance England* (Cambridge: Cambridge University Press, 1998).

Hard, G., *Spuren und Spurenleser: zur Theorie und Ästhetik des Spurenlesens in der Vegetation und anderswo* (Osnabrück: Universitätsverlag Rasch, 1995).

Harré, R., 'Recovering the Experiment', *Philosophy*, 73 (1998), pp. 353–77.

—, 'The Materiality of Instruments in a Metaphysics for Experiments', in H. Radder (ed.), *The Philosophy of Scientific Experimentation* (Pittsburgh, PA: University of Pittsburgh Press, 2003), pp. 19–38.

Hausman, J., and D. A. Wise, *Social Experimentation* (Chicago, IL: University of Chicago Press, 1985).

Heathcote, A. W., 'William Whewell's Philosophy of Science', *British Journal for the Philosophy of Science*, 4 (1954), pp. 302–14.

Heidelberger, M., 'Die Erweiterung der Wirklichkeit im Experiment', in M. Heidelberger and F. Steinle (eds), *Experimental Essays: Versuche zum Experiment* (Baden-Baden: Nomos, 1998), pp. 71–92.

—, and F. Steinle (eds), *Experimental Essays: Versuche zum Experiment* (Baden-Baden: Nomos, 1998).

Von Helmholtz, H., *Vorträge und Reden*, 2 vols (Braunschweig: Vieweg und Sohn, 1903).

Herrndorf, W., *Arbeit und Struktur* (Berlin: Rowohlt, 2013).

Hilpinen, R., 'On Experimental Questions', in D. Batens and J. P. v. Bendegem (eds), *Theory and Experiment: Recent Insights and New Perspectives on their Relation* (Dordrecht: Reidel, 1988), pp. 15–32.

Hobbes, T., *Leviathan, or, the Matter, Form, and Common Wealth, Ecclesiasticall and Civil* (London, 1651).

Höhler, S., and F. Luks (eds), *Beam us Up, Boulding! 40 Jahre 'Raumschiff Erde': Themenheft zum 40. Jubiläum von Kenneth E. Bouldings 'Operating Manual for Spaceship Earth' (1966)* (Hamburg: Vereinigung für Ökologische Ökonomie, 2006).

Holmes, F., *Eighteenth-Century Chemistry as an Investigative Enterprise* (Berkeley, CA: Office for History of Science and Technology, University of California at Berkeley, 1989).

Hughes, T. P., *Human-built World: How to Think about Technology and Culture* (Chicago, IL: University of Chicago Press, 2004).

Hume, D., *A Treatise of Human Nature: Being an Attempt to Introduce the Experimental Method of Reasoning into Moral Subjects*, 3 vols (London: Printed for John Noon, 1739–40).

Hutchinson, G. E., 'Circular Causal Systems in Ecology', *Annals of the New York Academy of Sciences*, 50 (1948), pp. 221–46.

Ingensiep, H. W., 'Brückenschläge – zur Sprache der Ökologie', in B. Busch (ed.), *Jetzt ist die Landschaft ein Katalog voller Wörter: Beiträge zur Sprache der Ökologie* (Göttingen: Wallstein, 2007), pp. 128–37.

—, and R. Hoppe-Sailer, *NaturStücke: Zur Kulturgeschichte der Natur*, (Ostfildern: Edition Tertium, 1996).

Jamison, A., *The Making of Green Knowledge* (Cambridge: Cambridge University Press, 2001).

Janich, P., 'Was macht experimentelle Resultate empiriehaltig? Die methodisch-kulturalistische Methode im Experiment', in M. Heidelberger and F. Steinle (eds), *Experimental Essays: Versuche zum Experiment* (Baden-Baden: Nomos, 1998), pp. 93–114.

Jardine, L., *Francis Bacon: Discovery and the Art of Discourse* (Cambridge: Cambridge University Press, 1974).

—, and M. Silverthorne, 'Introduction', in L. Jardine and M. Silverthorne (eds), *Francis Bacon, the New Organon* (Cambridge: Cambridge University Press, 2000), pp. vii–xxviii.

— (eds), *Francis Bacon, the New Organon* (Cambridge: Cambridge University Press, 2000).

Jardine, N., J. A. Secord and E. C. Spary, *Cultures of Natural History* (Cambridge: Cambridge University Press, 1996).

Jasanoff, S., *Designs on Nature: Science and Democracy in Europe and United States* (Princeton, NJ: Princeton University Press, 2005).

—, *States of Knowledge: The Co-Production of Science and the Social Order* (London: Routledge, 2006).

Jou, C., 'Counting Calories', *Chemical Heritage*, 29 (2011), pp. 27–31.

Kant, I., '*Beschluß von der Grenzbestimmung der reinen Vernunft §57*', in *Prolegomena zu einer jeden künftigen Metaphysik, die als Wissenschaft wird auftreten können*, ed. K. Vorländer (Leipzig: Felix Meiner, 1920), pp. 120–30.

Karn, B., and P. Aguar (organizers), 'Nanotechnology and Life Cycle Assessment', workshop report by W. Klöpffer (writing team coordinator) (Washington, DC: Woodrow Wilson International Center for Scholars, 2007).

Kitching, R., 'A Reflection of the Truth', in M. Canfield (ed.), *Field Notes on Science and Nature* (Cambridge, MA: Harvard University Press, 2011), pp. 67–87.

Klein, U., *Experiments, Models, Paper Tools* (Stanford, CA: Stanford University Press, 2003).

Knorr-Cetina, K., *The Manufacture of Knowledge: An Essay on the Constructivist and Contested Nature of Science* (Oxford: Pergamon Press, 1981).

—, 'The Couch, the Laboratory, and the Cathedrale: On the Relationship between Experiment and Laboratory in Science', in A. Pickering (ed.), *Science as Practice and Culture* (Chicago, IL: University of Chicago Press, 1992), pp. 113–38.

—, 'Die Manufaktur der Natur – oder: Die alterierten Naturen der Naturwissenschaft', in Institut für Wissenschafts- und Technikforschung (ed.), *Die 'Natur' der Natur* (Bielefeld: Institut für Wissenschafts- und Technikforschung, 1999), pp. 104–19.

—, *Epistemic Cultures: How the Sciences Make Knowledge* (Cambridge, MA: Harvard University Press, 1999).

Köchy, K., 'Lebewesen im Labor. Das Experiment in der Biologie', *Philosophia naturalis*, 43 (2006), pp. 74–110.

—, and G. Schiemann (eds), 'Natur im Labor', *Philosophia naturalis*, 43 (2006), pp. 1–9.

Kohler, R. E., *Landscapes and Labscapes: Exploring the Lab-field Frontier in Biology* (Chicago, IL: University of Chicago Press, 2002).

—, 'Lab History: Reflections', *Isis*, 99 (2008), pp. 761–8.

Kötter, R., 'Experiment, Simulation und orientierender Versuch: Anmerkungen zur Experimentalkultur der Biowissenschaften', in C. Brunold, P. Balsiger, J. B. Bucher and C. Körner (eds), *Wald und CO2: Ergebnisse eines ökologischen Modellversuchs* (Bern: Haupt, 2001), pp. 41–50.

Krohn, W., 'Einleitung', in W. Krohn (ed.), *Francis Bacon: Neues Organon: lateinisch-deutsch*, 2 vols (Hamburg: Meiner, 1990), vol. 1, pp. ix–lvi.

—, G. Böhme, W. van den Daele, R. Hohlfeld and W. Schäfer, *Finalization in Science: The Social Orientation of Scientific Progress* (Dordrecht: Reidel, 1983).

—, and W. van den Daele, 'Science as an Agent of Change: Finalization and Experimental Implementation', *Social Science Information*, 37 (1998), pp. 191–222.

—, and J. Weyer, 'Gesellschaft als Labor: Die Erzeugung sozialer Risiken durch experimentelle Forschung', *Soziale Welt*, 40 (1989), pp. 349–73.

—, 'Society as a Laboratory: The Social Risks of Experimental Research', *Science and Public Policy*, 21 (1994), pp. 173–83.

Krohs, U., 'Wissenschaftstheoretische Rekonstruktionen', in U. Krohs and G. Toepfer (eds), *Philosophie der Biologie* (Frankfurt am Main: Suhrkamp, 2005), pp. 304–21.

Kuhn, T. S., *The Structure of Scientific Revolutions* (Chicago, IL: University of Chicago Press, 1962).

—, *The Essential Tension: Selected Studies in Scientific Tradition and Change* (Chicago, IL: University of Chicago Press, 1977).

Kutschmann, W., *Der Naturwissenschaftler und sein Körper: Die Rolle der 'inneren Natur' in der experimentellen Naturwissenschaft der frühen Neuzeit* (Frankfurt am Main: Suhrkamp, 1986).

Kwa, C., *Styles of Knowing: A New History of Science from Ancient Times to the Present* (Pittsburgh, PA: University of Pittsburgh Press, 2011).

Lacey, H., 'On the Aims and Responsibilities of Science', *Principia*, 11 (2007), pp. 45–62.

—, 'The Life and Times of Transgenics', in B. Bensaude-Vincent, S. Loeve, A. Nordmann and A. Schwarz (eds), *Attractive Objects* (forthcoming).

Lachmund, J., *Greening Berlin: The Co-Production of Science, Politics, and Nature* (Cambridge, MA: MIT Press, 2012).

Lakatos, I., 'Falsification and the Methodology of Scientific Research Programmes', in I. Lakatos and A. Musgrave (eds), *Criticism and the Growth of Knowledge* (Cambridge: Cambridge University Press, 1970), pp. 91–195.

—, *The Methodology of Scientific Research Programmes*, Vol. 1, ed. J. Worral and G. Currie (Cambridge: Cambridge University Press, 1978).

Lange, R., *Experimentalwissenschaft Biologie. Methodische Grundlagen und Probleme einer technischen Wissenschaft vom Lebendigen* (Würzburg: Königshausen & Neumann, 1999).

Langenscheidts Taschenwörterbuch, *Lateinisch-Deutsch*, Vol. 1 (Berlin: Langenscheidt, 1963).

Latour, B., 'Give me a Laboratory and I Will Raise the World', in K. Knorr-Cetina and M. Mulkay (eds), *Science Observed* (London: Sage, 1983), pp. 141–70.

—, *Der Berliner Schlüssel: Erkundungen eines Liebhabers der Wissenschaften* (Berlin: Akademie Verlag, 1996).

—, 'From the World of Science to the World of Research?', *American Association for the Advancement of Science*, 280 (1998), pp. 208–9.

—, *Politiques de la nature: comment faire entrer les sciences en démocratie* (Paris: Editions La Découverte, 1999).

—, and S. Woolgar, *Laboratory Life: The Social Construction of Scientific Facts* (Beverly Hills, CA: Sage, 1979).

Laudan, L., 'William Whewell on the Consilience of Inductions', *Monist*, 55 (1971), pp. 368–91.

Laudan, R., 'Natural Alliance or Forced Marriage? Changing Relations between the Histories of Science and Technology', *Technology and Culture*, 36 (1995) (Supplement: Snapshots of a Discipline: Selected Proceedings from the Conference on Critical Problems and Research Frontiers in the History of Technology, Madison, Wisconsin, 30 October–3 November 1991), pp. 17–28.

Lavoisier, A., *Traité élémentaire de chimie* (Chicago, IL: Encyclopaedia Britannica, 1952).

LeGrand, H. (ed.), *Experimental Inquiries: Historical, Philosophical and Social Studies of Experimentation in Science* (Dordrecht: Kluwer, 1990), pp. 49–80.

Lewis, R., *Language, Mind and Nature: Artificial Languages in England from Bacon to Locke* (Cambridge: Cambridge University Press, 2007).

Lichtenberg, G. C., *Schriften und Briefe*, ed. W. Promies, 6 vols (Frankfurt am Main: Zweitausendeins, 1994).

Light, A., 'Restoring Ecological Citizenship', in B. A. Minteer and B. P. Taylor (eds), *Democracy and the Claims of Nature* (New York: Rowman and Littlefield, 2002), pp. 153–72.

—, and J. M. Smith (eds), *Philosophies of Place (Philosophy and Geography III)* (Lanham, MD: Rowman and Littlefied Publishers, 1998).

Limoges, C., 'The Development of the Muséum d'Histoire Naturelle of Paris, 1800–1914', in R. Fox and G. Weisz (eds), *The Organization of Science and Technology in France 1808–1914* (Cambridge: Cambridge University Press, 1980), pp. 211–40.

Lindeman, R. L., 'The Trophic-dynamic Aspect of Ecology', *Ecology*, 23 (1942), pp. 399–418.

Livingstone, D. L., *Putting Science in its Place: Geographies of Scientific Knowledge* (Chicago, IL: University of Chicago Press, 2003).

Longino, H. E., 'Theoretical Pluralism and the Scientific Study of Behaviour', in S. H. Kellert, H. E. Longino and C. K. Waters (eds), *Scientific Pluralism: Minnesota Studies in the Philosophy of Science* (Minneapolis, MN and London: University of Minnesota Press, 2006), pp. 102–31.

Lovelock, J., 'The Gaia Hypothesis', in P. Bunyard (ed.), *Gaia in Action: Science of the Living Earth* (Edinburgh: Floris, 1996), pp. 15–33.

Löw, M., *Raumsoziologie* (Frankfurt am Main: Suhrkamp, 2001).

Luce, J., and L. Giorgi, 'Knowledge Politics and Converging Technologies', *Innovation: The European Journal of Social Science Research*, 22 (2009), pp. 1–5.

Lynch, M., and S. Woolgar (eds), *Representation in Scientific Practice* (Dordrecht: Kluwer, 1990).

Mach, E., *Analysis of Sensations* (New York: Dover, 1959).

Malpas, J., 'Finding Place: Spatiality, Locality, and Subjectivity', in A. Light and J. M. Smith (eds), *Philosophies of Place* (Lanham, MD: Rowman and Littlefield, 1998), pp. 21–44.

Malthus, T. R., *An Essay on the Principle of Population, as it Affects the Future Improvement of Society* (London: Macmillan & Co., 1798).

Margalef, R., *Perspectives in Ecological Theory* (Chicago, IL: University of Chicago Press, 1968).

Martin, B. R., and P. Nightingale (eds), *The Political Economy of Science, Technology and Innovation* (Cheltenham: Edward Elgar Publishing, 2000).

Martinez, O., 'The Dynamics of Border Interaction: New Approaches to Border Analysis', in C. H. Shofield (ed.), *Global Boundaries* (London: Routledge, 1993), pp. 1–15.

Massey, D., *For Space* (London: Sage, 2005).

Mayr, E., *Das ist Biologie: Die Wissenschaft des Lebens* (Heidelberg: Spektrum, 2000).

McIntosh, R. P., 'Pluralism in Ecology', *Annual Review of Ecology and Systematics*, 18 (1987), pp. 321–41.

—, 'Concept and Terminology of Homogeneity and Heterogeneity in Ecology', in J. Kolasa and S. T. A. Pickett (eds), *Ecological Heterogeneity* (New York: Springer, 1991), pp. 24–46.

Mcnaghten, P., and J. Urry, *Contested Natures* (London: Sage, 1998).

Merchant, C., 'Secrets of Nature: The Bacon Debates Revisited', *History of Ideas*, 69 (2008), pp. 147–62.

Michener, W. K., K. L. Bildstein, A. McKee, R. R. Parmenter, W. W. Hargrove, D. McClearn and M. Stromberg, 'Biological Field Stations: Research Legacies and Sites for Serendipity', *BioScience*, 59 (2009), pp. 300–10.

Mill, J. S., *System der deductiven und inductiven Logik*, 2 vols, trans. J. Schiel (Braunschweig: Vieweg und Sohn, 1868).

Miller, S., 'Science Confronts the Law', in W. Bainbridge and M. Roco (eds), *Managing Nano-bio-info-cogno Innovations: Converging Technologies in Society* (Dordrecht: Springer, 2006), pp. 279–84.

Minteer, B. A., and B. P. Taylor (eds), *Democracy and the Claims of Nature* (New York: Rowman and Littlefield, 2002).

Mitchell, S., 'Dimensions of Scientific Law', *Philosophy of Science*, 67 (2000), pp. 242–65.

Mitman, G., *The State of Nature: Ecology, Community, and American Social Thought, 1900–1950* (Chicago, IL: University of Chicago Press, 1992).

Moreno, J., *Undue Risk: Secret State Experiments on Humans* (London: Routledge, 2000).

Morgan, M. S., 'Experiments without Material Intervention: Model Experiments, Virtual Experiments and Virtually Experiments', in H. Radder (ed.), *The Philosophy of Scientific Experimentation* (Pittsburgh, PA: University of Pittsburgh Press, 2003), pp. 216–35.

Mumford, L., 'Science as Technology', *Proceedings of the American Philosophical Society*, 105 (1961), pp. 506–11.

Muntersbjorn, M., 'Francis Bacon's Philosophy of Science: Machina intellectus and forma indita', *Philosophy of Science*, 70 (2002), pp. 1137–48.

Naess, A., 'The Shallow and the Deep, the Long-range Ecology Movement: A Summary', *Inquiry*, 16 (1973), pp. 95–100.

Naumann, E., 'Die Rohkultur des Heleoplanktons', in E. Abderhalden (ed.), *Handbuch der biologischen Arbeitsmethoden. Methoden der Süsswasserbiologie*, part 2 (Berlin: Urban und Schwarzenberg, 1925), p. 283.

Nordhaus, T., and M. Shellenberger, *Break Through: From the Death of Environmentalism to the Politics of Possibility* (Boston, MA: Houghton Mifflin Company, 2006).

Nordmann, A. (rapporteur), *Converging Technologies: Shaping the Future of European Societies* (Brussels: EC Report, 2004), at ec.europa.eu/research/conferences/2004/ntw/pdf/final_report_en.pdf [accessed 18 September 2013].

—, *Wittgenstein's Tractatus: An Introduction* (Cambridge: Cambridge University Press, 2005).

—, 'Collapse of Distance: Epistemic Strategies of Science Ando Technoscience', *Danish Yearbook of Philosophy*, 41 (2007), pp. 7–34.

—, 'European Experiments', *Osiris*, 24 (2009), pp. 278–302.

—, 'Philosophy of Science', in B. Clarke and M. Rossini (eds), *The Routledge Companion to Literature and Science* (London: Routledge, 2010), pp. 362–73.

—, and A. Schwarz, 'The Lure of the "Yes": The Seductive Power of Technoscience', in M. Kaiser, M. Kurath, S. Maasen and C. Rehmann-Sutter (eds), *Assessment Regimes of*

Technology: Regulation, Deliberation and Identity Politics of Nanotechnology (Dordrecht: Springer, 2010), pp. 255–77.

Norton, J. D., 'A Material Theory of Induction', *Philosophy of Science*, 70 (2003), pp. 647–70.

O'Neal, J. C., *The Authority of Experience: Sensationist Theory in the French Enlightenment* (University Park, PA: Pennsylvania State University Press, 1996).

Odum, E., 'The Mesocosm', *BioScience*, 24 (1984), pp. 558–62.

Ott, K., *Aufbruch und Wandel: Regelwerke für einen Green New Deal* (2010), at http://www.bfn.de/fileadmin/MDB/documents/ina/vortraege/2010_Sommerakademie-Konrad-Ott.pdf [accessed 4 December 2013].

Outram, D., 'New Spaces in Natural History', in N. Jardine, J. A. Secord and E. Spary (eds), *Cultures of Natural History* (Cambridge: Cambridge University Press, 1996), pp. 249–479.

Palmer, C. L., *Work at the Boundaries of Science: Information and the Interdisciplinary Research Process* (Dordrecht: Kluwer Academic Publishers, 2001).

Parsons, K. (ed.), *The Science Wars: Debating Scientific Knowledge and Technology* (New York: Prometheus Books, 2003).

Peirce, C. S., *Pragmatism and Pragmaticism: Collected Papers of Charles Sanders Peirce*, 8 vols (Cambridge, MA: Harvard University Press, 1934).

Perniola, M., 'Logique de la seduction', *Traverses*, 18 (1980), pp. 2–9.

Pesic, P., 'Wrestling with Proetus: Francis Bacon and the "Torture" of Nature', *Isis*, 90 (1999), pp. 81–94.

Pezzoli, K., 'Sustainable Development Literature: A Transdisciplinary Bibliography', *Journal of Environmental Planning and Management*, 40 (1997), pp. 575–601.

Pias, C. (ed.), *Cybernetics – Kybernetik: The Macy-Conferences (1946–1953)*, 2 vols (Zurich: Diaphanes, 2003).

Pickering, A. (ed.), *Science as Practice and Culture* (Chicago, IL: University of Chicago Press, 1991).

—, 'From Science as Knowledge to Science as Practice', in A. Pickering (ed.), *Science as Practice and Culture* (Chicago, IL: University of Chicago Press, 1992), pp. 1–29.

Pohl, C., and G. H. Hadorn (eds), *Principles for Designing Transdisciplinary Research* (Munich: Oekom-Verlag, 2007).

Pomata, G., 'Observation Rising: Birth of an Epistemic Genre, 1500–1650', in L. Daston and E. Lunbeck (eds), *Histories of Scientific Observation* (Chicago, IL: University of Chicago Press, 2011), pp. 45–80.

Poovey, M., *A History of the Modern Fact* (Chicago, IL: University of Chicago Press, 1998).

Popper, K., *The Open Society and its Enemies: The Spell of Plato* (Princeton, NJ: Princeton University Press, 1945).

—, 'Das Wachstum der wissenschaftlichen Erkenntnis', in D. Miller (ed.), *Karl Popper Lesebuch* (Tübingen: Mohr, 1995), pp. 154–63.

Porter, T. M., 'Objectivity and Authority: How French Engineers Reduced Public Utility to Numbers', *Poetics Today*, 2 (1991), pp. 245–65.

Potthast, T., '"Ökologie" als Brücke zwischen Wissen und Moral der Natur?', in B. Busch (ed.), *Jetzt ist die Landschaft ein Katalog voller Wörter: Beiträge zur Sprache der Ökologie* (Göttingen: Wallstein, 2007), pp. 138–45.

Prevos, P., *Origins of Modern Philosophy: The Problem of Induction* (May 2005), at http://www.prevos.net [accessed 4 February 2011].

Radder, H. (ed.), *The Philosophy of Scientific Experimentation* (Pittsburgh, PA: University of Pittsburgh Press, 2003).

Rader, K. A., *Making Mice: Standardizing Animals for American Biomedical Research, 1900–1955* (Princeton, NJ: Princeton University Press, 2004).

Radkau, J., 'Naturschutz und Nationalsozialismus – wo ist das Problem?', in J. Radkau and F. Uekötter (eds), *Naturschutz und Nationalsozialismus* (Frankfurt am Main: Campus, 2003), pp. 41–55.

—, *Die Ära der Ökologie: Eine Weltgeschichte* (Munich: Beck, 2011).

Rammert, W., 'Eine Soziologie, als ob Natur nicht zählen würde', at http://www.ts.tu-berlin.de/fileadmin/fg226/Rammert/articles/Soziologie_der_Natur.html [accessed 4 December 2013].

Rehmann-Sutter, C., 'An Introduction to Places', *Worldviews: Environment, Culture, Religion*, 2 (1998), pp. 171–7.

Reidy, M. S., G. Kroll and E. M. Conway, *Exploration and Science: Social Impact and Interaction* (Santa Barbara, CA: ABC-CLIO, 2007).

Reif, W.-E., 'Practice and Theory in Natural History: Domains and Epistemic Things', *Theory in Biosciences*, 118 (1999), pp. 161–74.

Rheinberger, H.-J., *Experiment, Differenz, Schrift* (Marburg: Basilisken-Presse, 1992).

—, 'Plädoyer für eine Wissenschaftsgeschichte des Experiments', *Theory in Biosciences*, 116 (1997), pp. 11–31.

— (ed.), *Conference 'the Shape of Experiment'*, Berlin, 2–5 June 2005, Preprint 318 (Berlin: MPI für Wissenschaftsgeschichte, 2005).

Riecken, H. W., *Social Experimentation: A Method for Planning and Evaluating Social Intervention* (New York: Academic Press, 1974).

Rip, A., 'Folk Theories of Nanotechnologists', *Science as Culture*, 15 (2006), pp. 349–65.

Roco, M. C., and W. S. Bainbridge, *Societal Implications of Nanoscience and Nanotechnology*, NSET Workshop Report (Arlington, VA: NSET, 2001).

—, B. Harthorn, D. Guston and P. Shapira, '*Innovative and Responsible Governance of Nanotechnology for Societal Development*', in M. C. Roco, C. Mirkin and M. Hersam (eds), *Nanotechnology Research Directions for Societal Needs in 2020* (Berlin: Springer, 2011), pp. 561–617.

Rose, G. D., 'Social Experiments in Innovative Environmental Management: The Emergence of Ecotechnology' (PhD thesis, University of Waterloo, 2003).

Rose, H., and S. Rose, *The Political Economy of Science: Ideology of/in the Natural Sciences* (London: Macmillan, 1976).

Rosenberg, A., *The Structure of Biological Science* (Cambridge: Cambridge University Press, 1985).

Roth, W.-H., and G. M. Bowen, 'Digitizing Lizards: The Topology of "Vision" in Ecological Fieldwork', *Social Studies of Science*, 29 (1999), pp. 719–64.

Röttgers, K., *Spuren der Macht: Begriffsgeschichte und Systematik* (Freiburg im Breisgau: Alber, 1990).

Rupke, N. (ed.), *Vivisection in Historical Perspective* (London: Routledge, 1990).

Russell, B., *History of Western Philosophy and its Connection with Political and Social Circumstances from the Earliest Times to the Present* (London: Allen and Unwin, 1946).

—, *The Problems of Philosophy* (Oxford: Oxford University Press, 1997).

Sabel, C. F., and J. Zeitlin, 'Learning from Difference: The New Architecture of Experimentalist Governance in the EU', *European Law Journal*, 13 (2007), pp. 271–327.

Schäfer, L., *Das Bacon-Projekt: Von der Erkenntnis, Nutzung und Schonung der Natur* (Frankfurt am Main: Suhrkamp, 1999).

Schaffer, S., 'The History and Geography of the Intellectual World: Whewell's Politics of Language', in M. Fisch and S. Schaffer (eds), *William Whewell: A Composite Portrait* (Oxford: Clarendon Press, 1991), pp. 201–32.

Schatzki, T. R., and K. Knorr-Cetina (eds), *The Practice Turn in Contemporary Theory* (London: Routledge, 2001).

Schepers, H., 'Heuristik, heuristisch', in *Historisches Wörterbuch der philosophischen Begriffe*, 13 vols (Basel: Schwabe, 1974), vol. 4, pp. 1115–20.

Schiebinger, L., 'Naming and Knowing: The Global Politics of Eighteenth-Century Botanical Nomenclatures', in P. Smith and B. Schmidt (eds), *Making Knowledge in Early Modern Europe: Practices, Objects, and Texts 1400–1800* (Chicago, IL: University of Chicago Press, 2007), pp. 90–105.

Schildknecht, C., 'Experiments with Metaphors: On the Connection between Scientific Method and Literary Form in Francis Bacon', in Z. Radman (ed.), *From a Metaphorical Point of View: A Multidisciplinary Approach to the Cognitive Content of Metaphor* (Berlin and New York: De Gruyter, 1995), pp. 27–50.

Schilpp, P. A., *The Philosophy of Karl Popper* (Chicago, IL: Open Court, 1974).

Schmid, H. B., *Moralische Integrität* (Frankfurt am Main: Suhrkamp, 2011).

Schmidt, J. C., 'What is a Problem? On Problem-oriented Interdisciplinarity', *Poiesis and Praxis*, 7 (2011), pp. 249–74.

Schmidt, K. F., *Green Nanotechnology: It's Easier than You Think* (Washington, DC: Woodrow Wilson International Center for Scholars, 2007).

Schöne, A., *Aufklärung aus dem Geist der Experimentalphysik. Lichtenbergsche Konjunktive* (Munich: C. H. Beck, 1982).

Schuler, R. M., 'Francis Bacon and Scientific Poetry', *Transactions of the American Philosophical Society*, 82 (1992), pp. 1–65.

Schwabe, G. H., 'August Thienemann in Memoriam', *Oikos*, 12 (1961), pp. 310–16.

Schwarte, L., *Philosophie der Architektur* (Munich: Fink, 2009).

Schwarz, A., 'Gestalten werden Systeme: Frühe Systemtheorie in der Ökologie', in K. Mathes, B. Breckling and K. Ekschmidt (eds), *Systemtheorie in der Ökologie* (Landsberg: Ekomed, 1996), pp. 35–45.

—, 'Modellierte Naturen und Raummodelle: Theoretische, ästhetische und strategische Eingriffe im Pariser Naturkundemuseum', *Berliner Schriften zur Museumskunde*, 22 (2006), pp. 155–64.

—, 'Wilde und liederliche Naturen: Stanislaw Lems nanotechnologische Vorbilder', in D. Korczak and A. Lerf (eds), *Zukunftspotentiale der Nanotechnologien: Erwartungen, Anwendungen, Auswirkungen* (Kröning: Asanger, 2007), pp. 103–26.

—, 'Baron Jakob von Uexküll: Das Experiment als Ordnungsprinzip in der Biologie', in A. Schwarz and A. Nordmann (eds), *Das bunte Gewand der Theorie* (Freiburg: Alber, 2009), pp. 207–34.

—, 'Escaping from Limits into Visions of Space', in A. Ferrari and S. Gammel (eds), *Visionen der Nanotechnologie* (Berlin: Akademische Verlagsgesellschaft, 2009), pp. 129–42.

—, 'Etymology and Original Sources of the Term "Ecology"', in A. Schwarz and K. Jax (eds), *Ecology Revisited: Reflecting on Concepts, Advancing Science* (Dordrecht: Springer, 2011), pp. 145–8.

—, 'In the Beginning, Man Created ...: Narrating the Drama of an Emerging Ecosystem', in B. Bensaude-Vincent, S. Loeve, A. Nordmann and A. Schwarz (eds), *Attractive Objects* (Pittsburgh, PA: Pittsburgh University Press, forthcoming).

—, and K. Jax, 'Early Ecology in the German-speaking World through WWII', in A. Schwarz and K. Jax (eds), *Ecology Revisited: Reflecting on Concepts, Advancing Science* (Dordrecht: Springer, 2011), pp. 231–76.

—, and W. Krohn, 'Experimenting with the Concept of Experiment: Probing the Epochal Break', in A. Nordmann, H. Radder and G. Schiemann (eds), *Science Transformed? Debating Claims of an Epochal Break* (Pittsburgh, PA: University of Pittsburgh Press, 2011), pp. 119–34.

—, and A. Nordmann, 'Unlimited Potential in a Limited World: The Political Economy of Eco- and Nanotechnologies', in M. Carrier and A. Nordmann (eds), *Science in the Context of Application: Methodological Change, Conceptual Transformation, Cultural Reorientation* (Berlin: Springer, 2010), pp. 317–36.

Serres, M., *Retour au contrat naturel* (Paris: Bibliothèque Nationale de France, 2000).

Shapiro, B. J., *A Culture of Fact: England, 1550–1720* (Ithaca, NY: Cornell University Press, 2000).

Shavit, A., and J. Griesemer, 'There and Back Again, or the Problem of Locality in Biodiversity Surveys', *Philosophy of Science*, 76 (2009), pp. 273–94.

Shrader-Frechette, K., and E. D. McCoy, *Method in Ecology: Strategies for Conservation* (Cambridge: Cambridge University Press, 1993).

—, 'Applied Ecology and the Logic of Case Studies', *Philosophy of Science*, 61 (1994), pp. 228–49.

Shofield, C. H. (ed.), *Global Boundaries* (London: Routledge, 1994).

Simon, H. A., 'Discovery, Invention, and Development: Human Creative Thinking', *Proc. Nat. Acad. Sci. USA*, 80 (1983), pp. 4569–71.

Small, A. W., 'The Future of Sociology', *Publications of the American Sociological Society*, 16 (1921), pp. 174–93.

Smith, P., *The Body of the Artisan: Art and Experience in the Scientific Revolution* (Chicago, IL: University of Chicago Press, 2004).

Snyder, L. J., *Reforming Philosophy: A Victorian Debate on Science and Society* (Chicago, IL: University of Chicago Press, 2006).

Society for the Diffusion of Useful Knowledge, 'An Account of Lord Bacon's *Novum organon scientarium* or, New Method of Studying the Sciences', in C. Knight (ed.), *Library of Useful Knowledge* (London: Baldwin and Cradock, 1828), pp. 1–40.

Stafford, B. M., *Visual Analogy: Consciousness as the Art of Connecting* (Cambridge, MA: MIT Press, 2001).

—, 'Thoughts Not Our Own: Whatever Happened to Selective Attention?', *Theory, Culture and Society*, 26 (2009), pp. 275–93.

Star, S. L., and J. R. Griesemer, 'Institutional Ecology, "Translations" and Boundary Objects: Amateurs and Professionals in Berkeley's Museum of Vertebrate Zoology, 1907–39', *Social Studies of Science*, 19 (1989), pp. 387–420.

Stavrakakis, Y., 'Passions of Identification: Discourse, Identification, and European Identity', in D. Howarth and J. Torfing (eds), *Discourse Theory in European Politics: Identity, Policy and Governance* (New York: Palgrave, 2005), pp. 68–92.

Steiner, G., *Dix raisons (possibles) à la tristesse de pensée*, ed. and trans. P.-E. Dauzat (Paris: Albin Michel).

Steinle, F., *Explorative Experimente: Ampère, Faraday und die Ursprünge der Elektrodynamik* (Wiesbaden: Franz Steiner, 2005).

—, 'Experiment', in F. Jäger (ed.), *Enzyklopädie der Neuzeit*, 3 vols (Stuttgart and Weimar: Metzler, 2006), vol. 3, pp. 722–8.

Stern, D. G., 'The Practical Turn', in S. P. Turner and P. A. Roth (eds), *The Blackwell Guide to the Philosophy of the Social Sciences* (Oxford: Blackwell, 2003), pp. 185–206.

Stokes, D., *Pasteur's Quadrant: Basic Science and Technological Innovation* (Washington, DC: Brookings Institution Press, 1997).

Sukopp, H., *Rückeroberung? Natur im Großstadtbereich* (Wien: Picus, 2003).

Thienemann, A., 'Zwecke und Ziele der Internationalen Vereinigung für theoretische und angewandte Limnologie', *Verhandlungen der Internationalen Vereinigung für theoretische und angewandte Limnologie*, 1 (1923), pp. 1–5.

—, 'Der Nahrungskreislauf im Wasser', *Zoologischer Anzeiger*, Suppl. 2 (1927), pp. 29–79.

—, *Die Bedeutung der Limnologie für die Kultur der Gegenwart* (Stuttgart: Schweizerbart'sche Verlagsbuchhandlung, 1935).

—, *Grundzüge einer allgemeinen Ökologie* (Stuttgart: Schweizerbart'sche Verlagsbuchhandlung, 1939).

—, 'Vom Wesen der Ökologie', *Biologia Generalis*, 3/4 (1942) (special edn), pp. 312–31.

—, 'Vom Gebrauch und vom Mißbrauch der Gewässer in einem Kulturlande', *Archiv für Hydrobiologie*, 45 (1951), pp. 557–83.

—, 'Wasser – Das Blut der Erde', in F. Oppenberg (ed.), *Handbuch der Schutzgemeinschaft Deutscher Wald: Uns ruft der Wald* (Rheinhausen: Verlagsanstalt Rheinhausen, 1954), pp. 45–9.

Tippett, J., '"Think like an Ecosystem": Embedding a Living System Paradigm into Participatory Planning', *Systemic Practice and Action Research*, 17 (2005), pp. 603–22.

Trepl, L., *Geschichte der Ökologie: Vom 17. Jahrhundert bis zur Gegenwart* (Frankfurt am Main: Athenäum, 1987).

Tweedale, G., '"Days at the Factories": A Tour of Victorian Industry with the Penny Magazine', *Technology and Culture*, 29 (1988), pp. 888–903.

UN Secretary General, *Progress to Date and Remaining Gaps in the Implementation of Outcomes of the Major Summits in the Area of Sustainable Development, as well as the Analysis of the Themes of the Conference*, Report for the Preparatory Committee for the United Nations Conference on Sustainable Development (New York: United Nations, 2010).

Urbach, P., *Francis Bacon's Philosophy of Science: An Account and a Reappraisal* (Chicago, IL: Open Court, 1987).

Valkenburg, G., 'Sustainable Technological Citizenship', *European Journal of Social Theory*, 15 (2012), pp. 471–87.

Valsangiacomo, A., *Die Natur der Ökologie* (Zurich: Hochschulverlag ETH Zürich, 1998).

Vernadsky, V. I., *La Biosphère* (Paris: Alcan, 1929).

Voigt, A., 'The Rise of Systems Theory in Ecology', in A. Schwarz and K. Jax (eds), *Ecology Revisited: Reflecting on Concepts, Advancing Science* (Dordrecht: Springer, 2011), pp. 183–94.

Walde, A., *Lateinisches etymologisches Wörterbuch* (Heidelberg: Carl Winter's Universitäts-Buchhandlung, 1939).

Walls, L. D., *Consilience Revisited*, electronic book review (December 1999), at http://electronicbookreview.com/thread/criticalecologies/biophilial [accessed 15 August 2011].

Wasmund, E., 'Wissenschaftsprovinzen', *Deutsche Rundschau*, 52 (1926), pp. 243–53.

Waters, K., 'How Practical Know-how Contextualizes Theoretical Knowledge: Exporting Causal Knowledge from Laboratory to Nature', *Philosophy of Science*, 75 (2008), pp. 707–19.

Waterton, C., and B. Wynne, 'Building the European Union: Science and the Cultural Dimensions of Environmental Policy', *Journal of European Public Policy*, 3 (1996), pp. 421–40.

Weber, M., *Philosophy of Experimental Biology* (Cambridge: Cambridge University Press, 2005).

Weigel, S., 'Das Gedankenexperiment: Nagelprobe auf die facultas fingendi in Wissenschaft und Literatur', in T. Macho and A. Wunschel (eds), *Science and Fiction: Über Gedankenexperimente in Wissenschaft, Philosophie und Literatur* (Frankfurt am Main: Fischer, 2004), pp. 183–208.

Weingart, P., W. Krohn and M. Carrier (eds), *Nachrichten aus der Wissensgesellschaft. Analysen zur Veränderung der Wissenschaft* (Weilerswist: Velbrück Wissenschaft, 2007).

Welzer, H., and H.-G. Soeffner (eds), *KlimaKulturen: Soziale Wirklichkeiten im Klimawandel* (Frankfurt am Main: Campus, 2010).

Whatmore, S., *Hybrid Geographies: Natures, Cultures, Spaces* (London: Sage, 2002).

Whewell, W., 'Preliminary Discourse on the Study of Natural Philosophy', *Quarterly Review*, 45 (1831), pp. 374–407.

—, 'Essay towards a First Approximation to a Map of Cotidal Lines', *Philosophical Transactions of the Royal Society of London*, 123 (1833), pp. 147–236.

—, 'Researches on the Tides: Fourth Series: On the Empirical Laws of the Tides in the Port of Liverpool', *Philosophical Transactions of the Royal Society of London*, 126 (1836), pp. 1–15.

—, *Aphorisms Concerning Ideas, Science, and the Language of Science* (La Vergne, TN: Kessinger Publishing, 1840).

—, *Philosophy of the Inductive Sciences: Founded upon their History*, 2 vols, 2nd edn (London: Parker and Son, 1847).

—, *Novum Organon Renovatum, being the Second Part of the Philosophy of the Inductive Sciences*, 3rd edn (London: Parker and Son, 1858).

—, *History of the Inductive Sciences*, 2 parts, 3rd edn (London: Parker and Son, 1857; repr. London: Frank Cass & Co. Ltd, 1967).

—, *Theory of Scientific Method*, ed. R. E. Butts (Indianapolis, IN: Hackett Publishing, 1989).

Wickson, F., 'Transdisciplinary Research: Characteristics, Quandaries and Qualities', *Nature*, 456 (2008), p. 29.

Wilson, E. O., *Consilience: The Unity of Knowledge* (New York: Alfred A. Knopf, 1998).

Wimsatt, W., *Re-engineering Philosophy for Limited Beings* (Cambridge, MA: Harvard University Press, 2007).

Windelband, W., *Präludien: Aufsätze und Reden zur Einleitung in die Philosophie* (Tübingen: Mohr, 1884).

—, *Geschichte und Naturwissenschaft: Rede zum Antritt des Rektorats der Kaiser-Wilhelm-Universität Strassburg; gehalten am 1. Mai 1894*, 2nd edn (Strasbourg: Heitz, 1900).

Wittgenstein, L., *Philosophical Investigations* (Oxford: Basil Blackwell, 1958).

Wolters, G., and J. G. Lennox (eds), *Concepts, Theories, and Rationality in the Biological Sciences* (Konstanz/Pittsburgh, PA: Universitätsverlag Konstanz/University of Pittsburgh Press, 1995).

Woods, B., 'Political Economy of Science', in G. Ritzer (ed.), *The Blackwell Encyclopedia of Sociology* (Oxford: Blackwell Publishing, 2007), pp. 3436–9.

Worster, D., *Nature's Economy: A History of Ecological Ideas* (Cambridge: Cambridge University Press, 1985).

Wynne, B., 'Risk and Environment as Legitimatory Discourses of Technology: Reflexivity Inside Out?', *Current Sociology*, 50 (2002), pp. 459–76.

—, *Risk and Reflexivity: Towards the Retrieval of Human Universals* (London: Sage Publications, 2008).

Yourcenar, M., *Fires*, trans. D. Katz (Chicago, IL: University of Chicago Press 1994).

Yuval, M., and J. Lezaun, 'Regulatory Experiments: Genetically Modified Crops and Financial Derivatives on Trial', *Science and Public Policy*, 33 (2006), pp. 179–90.

Zahar, E., 'Logic of Discovery or Psychology of Invention?', *British Journal for the Philosophy of Science*, 34 (1983), pp. 243–61.

Zedler, J. H., *Grosses vollständiges Universal-Lexicon aller Wissenschaften und Künste welche bisshero durch menschlichen Verstand und Witz erfunden und versbessert worden*, 64 vols (Halle and Leipzig, 1732–52).

Zittel, C., 'Introduction', in C. Zittel, G. Engel, R. Nanni and N. C. Karafyllis (eds), *Philosophies of Technology: Francis Bacon and his Contemporaries* (Leiden and Boston, MA: Brill, 2008), pp. 19–29.

—, G. Engel, R. Nanni and N. C. Karafyllis (eds), *Philosophies of Technology: Francis Bacon and his Contemporaries* (Leiden and Boston, MA: Brill, 2008).

INDEX

,